Elementary Fourier Optics for Science and Engineering Students

Online at: https://doi.org/10.1088/978-0-7503-6392-1

IOP Series in Advances in Optics, Photonics and Optoelectronics

SERIES EDITOR

Professor Rajpal S Sirohi Consultant Scientist

About the Editor

Rajpal S Sirohi is currently working as a faculty member in the Department of Physics, Alabama A&M University, Huntsville, AL, USA. Prior to this, he was a consultant scientist at the Indian Institute of Science, Bangalore, and before that he was Chair Professor in the Department of Physics, Tezpur University, Assam. During 2000–2011, he was an academic administrator, being vice-chancellor to a couple of universities and the director of the Indian Institute of Technology, Delhi. He is the recipient of many international and national awards and the author of more than 400 papers. Dr Sirohi is involved with research concerning optical metrology, optical instrumentation, holography, and the speckle phenomena.

About the series

Optics, photonics, and optoelectronics are enabling technologies in many branches of science, engineering, medicine, and agriculture. These technologies have reshaped our outlook and our ways of interacting with each other, and have brought people closer together. They help us to understand many phenomena better and provide deeper insight into the functioning of nature. Further, these technologies themselves are evolving at a rapid rate. Their applications encompass very large spatial scales, from nanometres to the astronomical scale, and a very large temporal range, from picoseconds to billions of years. This series on advances in optics, photonics, and optoelectronics aims to cover topics that are of interest to both academia and industry. Some of the topics to be covered by the books in this series include biophotonics and medical imaging, devices, electromagnetics, fibre optics, information storage, instrumentation, light sources, charge-coupled devices (CCDs) and complementary metal oxide semiconductor (CMOS) imagers, metamaterials, optical metrology, optical networks, photovoltaics, free-form optics and its evaluation, singular optics, cryptography, and sensors.

About IOP ebooks

The authors are encouraged to take advantage of the features made possible by electronic publication to enhance the reader experience through the use of color, animation, and video and by incorporating supplementary files in their work.

A list of titles published in this series can be found here: https://iopscience.iop.org/bookListInfo/series-on-advances-in-optics-photonics-and-optoelectronics.

Elementary Fourier Optics for Science and Engineering Students

Hélène Ollivier

École Nationale Supérieure des Sciences Appliquées et de Technologie, CNRS, Institut FOTON-UMR 6082, Université de Rennes, 6 rue de Kerampont, 22300 Lannion, France

Osvaldo de Melo

Escuela Superior de Física y Matemáticas, Escuela Superior de Ingeniería Mecánica y Eléctrica, Instituto Politécnico Nacional UPALM, Ciudad de México, México

IOP Publishing, Bristol, UK

ISBN   978-0-7503-6392-1 (ebook)
ISBN   978-0-7503-6388-4 (print)
ISBN   978-0-7503-6389-1 (myPrint)
ISBN   978-0-7503-6391-4 (mobi)

DOI   10.1088/978-0-7503-6392-1

Multimedia content is available for this book from https://doi.org/10.1088/978-0-7503-6392-1.

Version: 20250801

IOP ebooks

British Library Cataloguing-in-Publication Data: A catalogue record for this book is available from the British Library.

Published by IOP Publishing, wholly owned by The Institute of Physics, London

IOP Publishing, No.2 The Distillery, Glassfields, Avon Street, Bristol, BS2 0GR, UK

US Office: IOP Publishing, Inc., 190 North Independence Mall West, Suite 601, Philadelphia, PA 19106, USA

From Hélène. To my sweet Sophie, Mila, and baby Louise, who joined the equation while this book was being written.

From Osvaldo. To Maruchy and Claudia: All the light in this book is yours.

Contents

Preface

Les mathématiques comparent les phénomènes les plus divers et découvrent les analogies secrètes qui les unissent.

Joseph Fourier

Fourier's theorem is not only one of the most beautiful results of modern analysis, but it may be said to furnish an indispensable instrument in the treatment of nearly every recondite question in modern physics.

Lord Kelvin

The interplay between mathematics and optics has profoundly shaped the way mankind perceives and manipulates light. When Jean Baptiste Joseph Fourier (1768–1830) introduced the concept of decomposing complex patterns into harmonic functions in 1807, he unwittingly laid the foundation for future breakthroughs in optical science. This mathematical innovation was further advanced in 1873, when Ernst Abbe (1840–1905) pioneered work on microscope resolution, establishing a direct link between the principles of diffraction and Fourier's transform, bridging the abstract mathematical domain with tangible optical applications.

The 20th century witnessed a rapid and transformative synergy between these two fields. The development of interferometry by Albert Abraham Michelson (1852–1931), who received the Nobel Prize in 1907, marked a turning point in the practical application of Fourier's theory to optical analysis, revolutionizing the way light was studied and measured. This momentum continued in the middle of the century with the work of André Maréchal (1916–2007) on the analysis of the combined effects of diffraction and aberrations. These milestones laid the foundation for cutting-edge innovations such as holography (for which Dennis Gabor received a Nobel Prize in 1971), computational imaging, and holographic data storage—technologies that underpin many aspects of modern science and engineering.

This book, *Elementary Fourier Optics for Science and Engineering Students*, was born out of a shared commitment to make the field of Fourier optics accessible, engaging and comprehensive for university students. Designed as a core text for science and engineering courses, it bridges the gap between the abstract mathematics underlying Fourier analysis and its wide-ranging applications in optical systems and technologies.

Today, Fourier optics stands as a cornerstone of modern optical science, providing a fundamental framework that combines the abstract elegance of mathematical principles with real-world engineering and applications. From improving the resolution of microscopes to enabling the creation of advanced holographic displays, Fourier optics has become an indispensable tool across many disciplines, driving advances in imaging, signal processing, and beyond.

Elementary Fourier Optics for Science and Engineering Students is designed to serve as both an introduction and a guide to this remarkable field. It bridges the gap between accessibility and rigour, ensuring that readers gain theoretical knowledge while exploring the depth and complexity of the subject. Whether you are new to Fourier optics or looking for a concise yet comprehensive reference, this book has been designed to demystify the subject while retaining its depth and complexity.

The development of this book has been a collaborative journey, driven by a shared vision to provide readers with both theoretical foundations and practical insights. We wanted to create a resource that would appeal to learners of all backgrounds, taking them step by step from the basics to more advanced topics.

The book is structured in such a way that each chapter builds on the previous one, providing clarity without sacrificing rigour, and ensuring that readers gain not only knowledge but also the confidence to apply Fourier optics in academic and professional contexts.

The structure of this book has been carefully crafted to ensure clarity while providing a comprehensive understanding of Fourier optics. It begins by equipping the reader with a robust mathematical toolkit, including essential concepts such as harmonic, delta, sinc, Bessel, and complex variable functions, in chapter 1. A concise review of geometric and wave optics provides a solid foundation with classic topics on interference and diffraction in chapter 2.

The focus then shifts to exploring the 1D Fourier transform in chapter 3. Readers will delve into Fourier series, fundamental theorems, and transforms of key functions such as the delta, comb, square, and trigonometric functions. Hands-on examples, using the open-source software ImageJ, make complex ideas tangible and encourage practical engagement. Building on this, chapter 4 introduces the 2D Fourier transform, covering topics such as the Fourier–Bessel transform, spatial frequency analysis, and image processing applications, again using ImageJ for demonstrations.

Chapter 5 introduces linear shift-invariant systems, point spread functions, and convolution. These concepts are linked to imaging and optical system design, providing a practical lens on theoretical principles. Diffraction phenomena are revisited in the context of Fourier analysis in chapter 6. Discussions of correlation, optical transfer functions, Parseval's theorem, and the angular spectrum of plane waves add depth to the treatment of the subject in chapter 7.

Finally, the book concludes with advanced topics in chapter 8 that demonstrate the broader applicability of Fourier optics. These include holography, Fourier transform infrared spectroscopy (FTIR), and computational tools such as discrete and fast Fourier transforms (FFT). The Whittaker–Shannon sampling theorem is introduced, linking classical ideas to modern industrial and research applications. Each chapter is complete with end-of-chapter problems, some of which include solutions.

Using a structured and engaging approach, *Elementary Fourier Optics for Science and Engineering Students* ensures that readers acquire both the analytical skills and the confidence needed to tackle complex optical problems in academic and professional settings.

Hélène Ollivier
Osvaldo de Melo

Acknowledgements

From Hélène: I would like to thank Thierry Chartier, a colleague who has always been generous and enthusiastic in our discussions about Fourier optics. I am grateful to my significant other, Stephen Wein, for patiently proofreading the drafts of my chapters. I thank my parents for passing on to me a deep curiosity and love for science. And finally, I thank my daughters, whose unique way of seeing the world continues to inspire me.

From Osvaldo: I want to thank my students Ricardo Domínguez and Elías Cervantes for their invaluable help in countless ways.

We are deeply grateful to IOP Publishing for their support in bringing this project to fruition. Their commitment to advancing knowledge has enabled us to share our passion for Fourier optics with a wider global audience.

We would like to thank our readers for joining us on this journey. Whether you are new to Fourier optics or looking for a concise reference, we hope that this book will serve as a valuable tool and inspire new perspectives in your studies and professional endeavours.

About the authors

Hélène Ollivier

Hélène Ollivier simultaneously obtained in 2017 her engineer diploma from Institut d'Optique Graduate School (Palaiseau, France), completed her master's degree in lasers, optics and matter, and graduated from Ecole Normale Supérieure de Paris-Saclay as a normalien student. She completed her doctoral research in quantum photonics and cavity quantum electrodynamics in the Optics and Semiconductors nanoStructures Group under the supervision of Professor Pascale Senellart, a CNRS research director at the Center for Nanosciences and Nanotechnologies (C2N). She developed and experimentally verified models that enable a precise optical characterization of single-photon sources based on quantum dots, which she used to conduct systematic studies of many sources fabricated at C2N. She successfully demonstrated experimental control over the symmetry of quantum dots and used finite element simulations to gain a deeper understanding of the involved mechanisms. These studies led to three first-author publications, one in *ACS Photonics* and two in *Physical Review Letters*. She has also co-authored seven additional papers published in other top journals such as *Nature Photonics*, *Nature Communications* and *Optica*.

As a laureate of the 2016 physics agrégation, a highly competitive examination for the Education Nationale in France, she is deeply involved in physics education. She taught high school physics and chemistry in the Paris region in 2021–22. From 2022 to 2024, she was responsible for the second year of a master's-level apprenticeship programme in an engineering school in Brittany, France (Ecole Supérieure d'Ingénieurs de Rennes), in which she has developed labworks and taught new courses on optical, electric, dielectric, and magnetic properties of materials, applied mathematics, fluid mechanics, object oriented programming and artificial intelligence at the masters level.

She has held a permanent position as Associate Professor at the Foton Institute (University of Rennes) since September 2024. She currently teaches subjects related to photonics at ENSSAT (École Nationale Supérieure des Sciences Appliquées et de Technologie) in Lannion and conducts research on room-temperature single-photon sources based on rare-earth-ion-doped optical fibres.

Osvaldo de Melo

Osvaldo de Melo is a distinguished physicist and educator specializing in semiconductor materials, thin-film growth, and nanotechnology. Born in Havana, Cuba in 1957, he earned his PhD in Physical Sciences from the University of Havana in 1992, focussing on ternary single crystal solutions of II–VI telluride compounds.

A dedicated educator with over 40 years of experience, he served as Full Professor of Physics at the University of Havana (1993–2021), Dean of the Faculty of Physics (2001–06), and visiting scholar at prestigious institutions including the Autonomous University of Madrid, the Instituto Politécnico Nacional (Mexico), and the Universidade Federal de Minas Gerais (Brazil). His extensive teaching portfolio encompasses courses at all academic levels, covering thermodynamics, optics, semiconductor physics, optoelectronics, spectroscopies, and diffraction. He has mentored generations of students across bachelor's, master's, and PhD programmes and organized major international scientific meetings such as the Latin American Symposium on Solid-State Physics and the Latin American Congress on Surface Science and Applications.

His research, documented in more than 100 peer-reviewed scientific publications, spans materials preparation, nanostructured semiconductors, and 2D layered systems including molybdenum oxides and Bi_2Te_3. With an h-index of 20 (Google Scholar) and more than 1400 citations, his scientific contributions have earned numerous honours including Cuba's National Physics Prize (2016) and multiple awards from the Cuban Academy of Sciences for breakthroughs across various areas of Physics and Materials Science.

As President of the Cuban Physical Society (2005–11), he championed international collaboration and science outreach. Currently working at both the Escuela Superior de Física y Matemáticas and the Escuela Superior de Ingeniería Mecánica y Eléctrica at Mexico's Instituto Politécnico Nacional (IPN), he investigates transition-metal oxides and dichalcogenides for next-generation electronic devices. Beyond academia, Dr de Melo is a passionate science communicator, having published over 40 popular articles on diverse topics in physics and related subjects.

IOP Publishing

Elementary Fourier Optics for Science and Engineering Students

Hélène Ollivier and Osvaldo de Melo

Chapter 1

Mathematical functions—a review

This chapter is dedicated to introducing the mathematical fundamentals necessary for a full understanding of the next chapters. It is designed to make the book self-contained. We begin with an overview of mathematical functions in one dimension. We then present some tools to manipulate them: scaling and translation, which will be useful in later chapters to mathematically describe optical objects in space for example. We explore trigonometric functions and identities, that are extensively used in optics. Next, we introduce key calculus concepts: derivation and integration, followed by examples of applications that will be referred to later in the book. After that, we focus on functions in two dimensions, as optical images are most often bidimensional. We then describe harmonic and special functions (the sinc function, Bessel functions and Gaussian function) which commonly appear in Fourier optics. Finally, we proceed to complex numbers, and further complex variable functions, which form the mathematical foundation of Fourier analysis.

1.1 Introduction to mathematical functions

In this section, we will remind the reader what mathematical functions are, and how to manipulate them. Some useful examples in the context of optics will be presented.

1.1.1 Definition

A mathematical function is an object that associates a value with another. This concept is useful in many fields and can be used to describe a dependency of some variable on another. As examples, we can think of the evolution of housing prices with time, or the evolution of temperature in space. In the context of optics, the two main varying parameters are space and time. The values of interest that vary with these parameters can be electric field amplitude, light intensity, contrast, refraction index, etc. All these are presented later in the book.

doi:10.1088/978-0-7503-6392-1ch1

Formally, a function f is written as:

$$\mathbb{R} \longrightarrow \mathbb{R}$$
$$x \longmapsto f(x)$$

The first line of the definition indicates that the object (number that the function takes as an argument) and the image (number that the function associates with the object) are both part of the set of all real numbers $\mathbb{R}$.

1.1.2 Examples

1.1.2.1 Linear functions
Very common examples of functions are linear functions, of the type:

$$\mathbb{R} \longrightarrow \mathbb{R}$$
$$x \longmapsto ax + b \text{ with } a, \ b \ \in \mathbb{R}$$

Such a function can be represented graphically on a bidimensional plot, see figure 1.1(a).

For example, the distance travelled by somebody walking at constant speed v varies linearly with time. In that case, the covered distance $d(t)$ varies with time as $d(t) = v \, t$. This is the function we just defined, with the coefficients a and b being equal to, respectively, v, the person's speed, and 0 m, the person's initial travelled distance. The distance covered by a walking person as a function of time is plotted in figure 1.1(b), for a constant speed equal to 1.5 m s^{-1}.

1.1.2.2 Quadratic functions
Quadratic functions have the following definition:

$$\mathbb{R} \longrightarrow \mathbb{R}$$
$$x \longmapsto \alpha_2 x^2 + \alpha_1 x + \alpha_0 \text{ with } \alpha_2, \ \alpha_1 \text{ and } \alpha_0 \in \mathbb{R}$$

Such a function can be represented graphically on a bidimensional plot as well, see figure 1.2(a).

For example, the altitude of a ball dropped from the origin of space and without initial speed in a field with gravity g follows a quadratic evolution with time: its

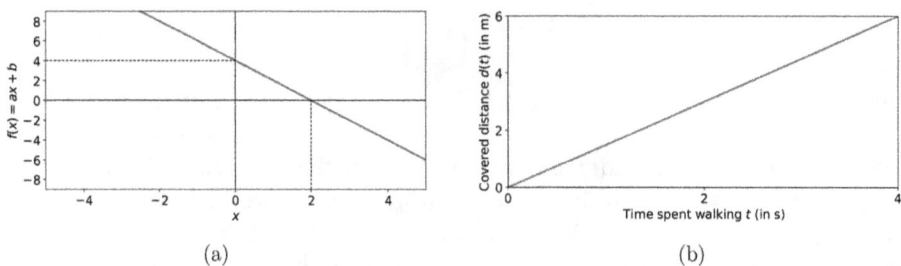

(a)   (b)

Figure 1.1. Graphic representation of linear functions (a) $f(x) = ax + b$ as a function of x, with coefficients $a = -2$ and $b = 4$, (b) evolution with time of the distance travelled by a person walking at constant speed $v = 1.5$ m s^{-1}.

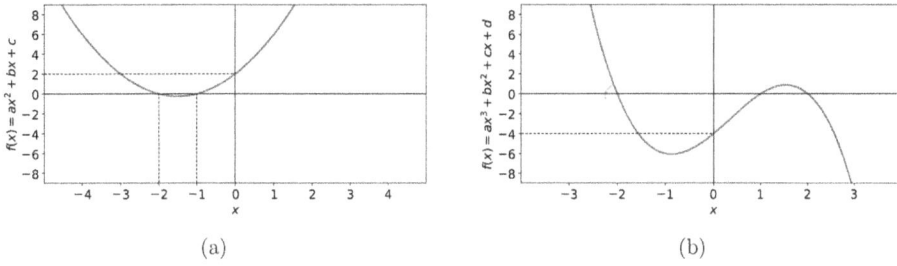

Figure 1.2. Graphic representation of polynomial functions (a) of degree 2 (quadratic function) of equation $f(x) = ax^2 + bx + c$ as a function of x, with coefficients $a = 1$, $b = 3$ and $c = 2$, (b) of degree 3 (cubic function) of equation $f(x) = ax^3 + bx^2 + cx + d$, with $a = -1$, $b = 1$, $c = 4$, and $d = -4$.

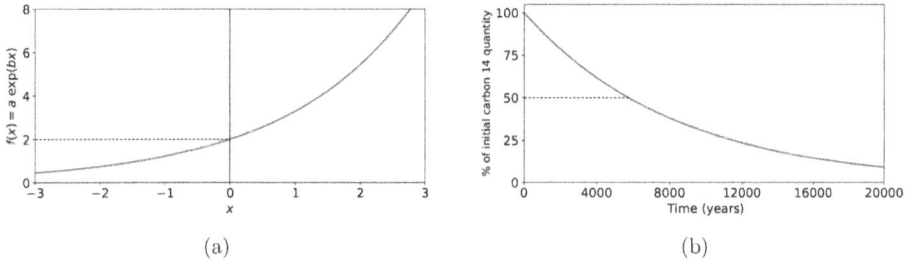

Figure 1.3. Graphic representation of exponential functions of equation $f(x) = a \exp(bx)$ as a function of x, (a) with coefficients $a = 2$ and $b = 0.5$, (b) with coefficients $a = 100\%$ and b was calculated from the half-life time of carbon 14.

distance from the origin is equal to $1/2gt^2$, which is equal to the expression of the function we just defined with $\alpha_2 = g/2$, $\alpha_1 = 0$ m s^{-1} (the initial speed of the ball) and $\alpha_0 = 0$ m (its initial position). Quadratic functions are part of a bigger ensemble of functions, that are the polynomial functions, that can be written under the following general form:

$$\mathbb{R} \longrightarrow \mathbb{R}$$
$$x \longmapsto \alpha_n x^n + \alpha_{n-1}x^{n-1} + \cdots + \alpha_1 x + \alpha_0 \text{ with } \alpha_n, \ldots, \alpha_0 \in \mathbb{R} \text{ and } n \in \mathbb{N}$$

where $\mathbb{N}$ refers to the ensemble of positive integer numbers. Such a function can be represented graphically on a bidimensional plot, see figure 1.2(b).

1.1.2.3 Exponential functions
Exponential functions are written as follows:

$$\mathbb{R} \longrightarrow \mathbb{R}$$
$$x \longmapsto \alpha \exp(\beta x) \text{ with } \alpha \text{ and } \beta \in \mathbb{R}$$

Such a function can be represented graphically on a bidimensional plot as well, see figure 1.3(a).

A typical example of exponential evolution is that of entities among which the number of disappearing or appearing entities per unit of time is proportional to their total number. We can think of the demographic evolution of a population with time, or a quantity of unstable radioactive material, etc. For example, the isotope of carbon having 14 nucleons exists in living organisms in a known proportion, most of the carbon having 12 nucleons. Carbon 14 is unstable and decays into nitrogen by emitting an extra particle, with a half-life of about 5700 years. That means that it takes 5700 years to see a given amount of carbon 14 decrease by a factor 2. The representation of such an evolution is shown in figure 1.3(b). Thus, knowing the proportion of carbon 14 that a former living organism contains makes it possible to date that organism's death. This datation technique was proposed by W F Libby, who won the Nobel Prize for it in 1960.

Some useful properties of exponential are the following:

$$\exp(a + b) = \exp(a)\exp(b)$$
$$\exp(a - b) = \exp(a)/\exp(b).$$

1.1.3 Scaling and translation

For this whole paragraph, let f be a function defined as:

$$\mathbb{R} \longrightarrow \mathbb{R}$$
$$x \longmapsto f(x)$$

Such a function can be represented graphically on a bidimensional plot such as the one shown in figure 1.4(a). We can use this function to define any other function that is scaled vertically or horizontally, or shifted toward any direction. The ability to scale and shift functions is a useful concept in the study of functions, as it allows one to decompose them as combinations of basic functions.

• To shift a function left (respectively right), one has to add (respectively substract) a shifting amount to the function argument. Mathematically, this can be written:

$$\mathbb{R} \longrightarrow \mathbb{R}$$
$$x \longmapsto f(x \pm b) \text{ with } b \in \mathbb{R} \text{ the shifting amount.}$$

These operations are represented in figures 1.4(b) and 1.4(c).

• To shift a function up (respectively down), one has to add (respectively substract) a shifting amount to the function itself. Mathematically, this can be written:

$$\mathbb{R} \longrightarrow \mathbb{R}$$
$$x \longmapsto f(x) \pm a \text{ with } a \in \mathbb{R} \text{ the shifting amount.}$$

These operations are represented in figures 1.4(d) and 1.4(e).

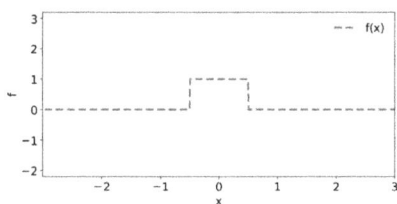

(a) Graphical representation of a rectangle function.

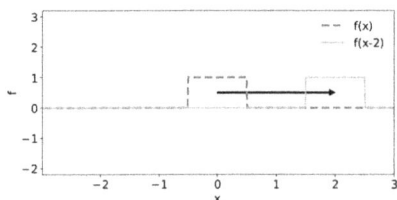

(b) Rectangle function from panel (a) translated to the right by 2 units.

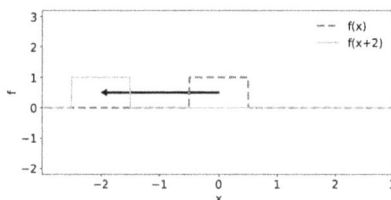

(c) Rectangle function from panel (a) translated to the left by 2 units.

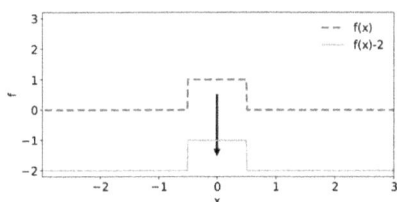

(d) Rectangle function from panel (a) translated downwards by 2 units.

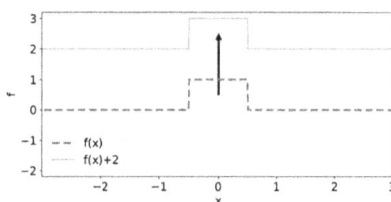

(e) Rectangle function from panel (a) translated upwards by 2 units.

Figure 1.4. Shifting operation on a rectangle function.

• To scale a function up (respectively down) horizontally, one has to multiply the argument of the function by a scaling amount smaller than 1 (respectively greater than 1). Mathematically, this can be written:

$$\mathbb{R} \longrightarrow \mathbb{R}$$
$$x \longmapsto f(\beta x) \text{ with } \beta \text{ the scaling factor.}$$

These operations are represented in figures 1.5(a) and 1.5(b).

• To scale a function up (respectively down) vertically, one has to multiply the function itself by a scaling amount greater than 1 (respectively smaller than 1). Mathematically, this can be written:

$$\mathbb{R} \longrightarrow \mathbb{R}$$
$$x \longmapsto \alpha f(x) \text{ with } \alpha \text{ the scaling factor.}$$

These operations are represented in figures 1.5(c) and 1.5(d).

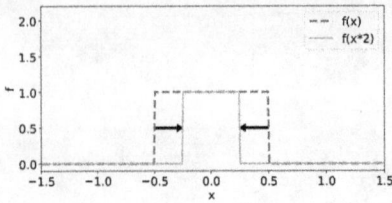

(a) Rectangle function from figure 1.4(a) scaled horizontally by a factor 1/2.

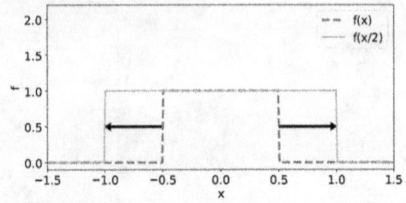

(b) Rectangle function from figure 1.4(a) scaled horizontally by a factor 2.

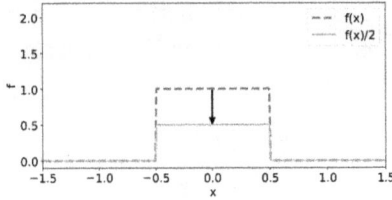

(c) Rectangle function from figure 1.4(a) scaled vertically by a factor 1/2.

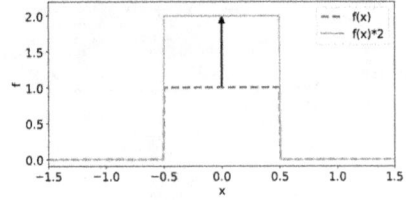

(d) Rectangle function from figure 1.4(a) scaled vertically by a factor 2.

Figure 1.5. Scaling operations on a rectangle function.

1.2 Trigonometry

Life is full of periodic events, and there exists a family of periodic events that are of great interest in science, since they basically rule our lives: waves. A wave is a periodic phenomenon that moves energy without moving matter. The sounds we hear are all waves, in the shape of compression and dilatation of air slices (this is why there is no sound transfer in a vacuum). The sea waves we play in at the beach are obviously waves as well, in the shape of variation of sea level. The last example we will give here is, of course, light, which is a spatio-temporal variation of the electromagnetic field. This last example does not need a medium to travel: this is why we can see stars although there is no material between the stars' atmosphere and ours.

1.2.1 Trigonometric functions

The building blocks to mathematically describe any periodic phenomenon are sinusoidal functions: sine and cosine, for which graphic representations are shown in figure 1.6.

Note that sine is an odd function ($\sin(-\theta) = -\sin(\theta)$) and cosine is an even function ($\cos(-\theta) = \cos(\theta)$). Sometimes in the calculations of this book, the ratio of these two functions will appear. It is then useful to introduce the tangent function, defined as:

$$\tan \theta = \frac{\sin \theta}{\cos \theta}.$$

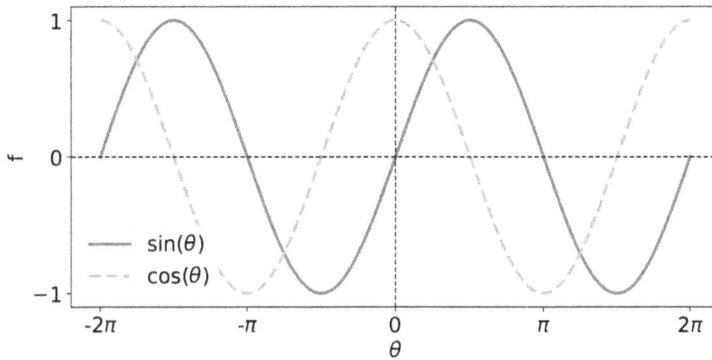

Figure 1.6. Graphic representations of sine and cosine, as the varying argument is represented in radians.

1.2.2 Trigonometric circle

A common and practical way to visualize trigonometric values and relationships between cosine and sine is to use the trigonometric circle, displayed in figure 1.7. This circle has a unitary radius. The lengths of the acute sides of the grey triangle are given by $\cos\theta$ and $\sin\theta$. As the angle parameter θ increases, $\cos\theta$ and $\sin\theta$ periodically sweep the $[-1, +1]$ interval.

1.2.3 Trigonometric formulas

From the trigonometric circle, we can see the following relationships:

$$\sin\left(\theta + \frac{\pi}{2}\right) = \cos\theta$$

$$\cos\left(\theta + \frac{\pi}{2}\right) = -\sin\theta$$

In order to be able to manipulate trigonometric functions, some usual formulas are to be known:

$$\cos(a + b) = \cos a \cos b - \sin a \sin b \qquad (1.1)$$

$$\sin(a + b) = \cos a \sin b + \sin a \cos b \qquad (1.2)$$

From which we can retrieve, by replacing $b \longrightarrow -b$:

$$\cos(a - b) = \cos a \cos b + \sin a \sin b \qquad (1.3)$$

$$\sin(a - b) = -\cos a \sin b + \sin a \cos b \qquad (1.4)$$

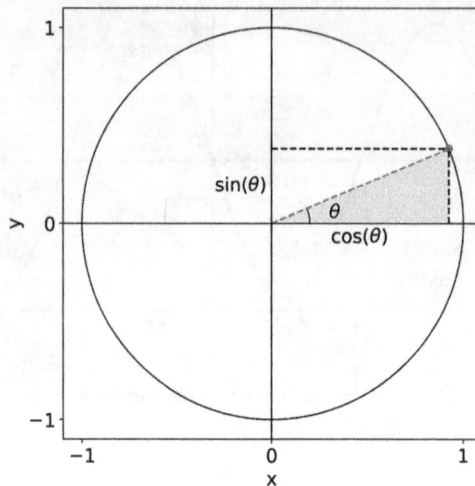

Figure 1.7. Trigonometric circle.

By doing linear combinations of the last sets of equations, we can retrieve three more useful formulas:

$$\cos a \cos b \; = \; \frac{1}{2}(\cos(a + b) + \cos(a - b))$$

$$\sin a \sin b \; = \; \frac{1}{2}(\cos(a - b) - \cos(a + b))$$

$$\sin a \cos b \; = \; \frac{1}{2}(\sin(a + b) + \sin(a - b))$$

From that last set of equations, we can retrieve, by setting $a = (p + q)/2$ and $b = (p - q)/2$, four more useful formulas:

$$\cos p + \cos q \; = \; 2\cos\left(\frac{p + q}{2}\right)\cos\left(\frac{p - q}{2}\right)$$

$$\cos p - \cos q \; = \; -2\sin\left(\frac{p + q}{2}\right)\sin\left(\frac{p - q}{2}\right)$$

$$\sin p + \sin q \; = \; 2\sin\left(\frac{p + q}{2}\right)\cos\left(\frac{p - q}{2}\right)$$

$$\sin p - \sin q \; = \; 2\cos\left(\frac{p + q}{2}\right)\sin\left(\frac{p - q}{2}\right)$$

Another interesting relationship that exists in trigonometry is the following:

$$\cos^2 x + \sin^2 x = 1.$$

1.3 Calculus

1.3.1 Derivation

1.3.1.1 Definition

The notion of derivative of a function is fundamental and abundantly used in this book. We limit ourselves to a brief definition: the derivative of a function f with respect to a variable x at abscissa x is written df/dx and is equal to:

$$\frac{df}{dx} = \lim_{dx \to 0} \frac{f(x + dx) - f(x)}{dx} \tag{1.5}$$

It is equal to the slope of the representative curve of f at abscissa x.

An important relationship between the sine and cosine functions (presented in section 1.2.1) is that they are proportional to each other's derivative, as follows:

$$\frac{d\cos\theta}{d\theta} = -\sin\theta$$

$$\frac{d\sin\theta}{d\theta} = \cos\theta$$

1.3.1.2 Use example: Taylor series

When considering small angles, which is a common approximation in optics and in physics more generally, it is possible to use Taylor series of sine and cosine around zero, which often simplifies things. The general formula for a Taylor series is the following:

$$f(x \simeq a) = \sum_{k=0}^{\infty} \frac{1}{k!} \frac{\partial^k f}{\partial x^k}\bigg|_{x=a} (x - a)^k \tag{1.6}$$

where the symbol ∂ refers to a derivative with respect to a given variable while the potential other variables on which the function depends are kept constant. For sine and cosine, if we limit ourselves to the second order term, we get:

$$\sin\theta \underset{\theta \ll 1}{\simeq} \theta$$

$$\cos\theta \underset{\theta \ll 1}{\simeq} 1 - \frac{\theta^2}{2}. \tag{1.7}$$

1.3.2 Integration

1.3.2.1 Definition

The integral of a function is defined as the area between its curve and the abscissa axis. It is thus defined between two bounds, let us say a and b. In that case, the integral is written:

$$\int_a^b f(x)\,dx = F(b) - F(a) = [F(x)]_a^b$$

where F is a function whose derivative is f.

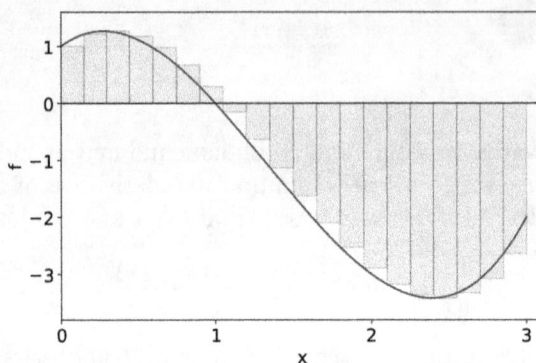

Figure 1.8. Illustration of the integral as the limit of the sum of rectangle areas.

To give an intuitive understanding of the mathematical expression of an integral, we can use Riemann's definition:

$$\int_a^b f(x)\,\mathrm{d}x = \lim_{\nu \to 0} \sum_{n=E(a/\nu)}^{n=E((b-a)/\nu)} f(n\nu)\nu$$

where $E(x)$ refers to the floor of x (first integer lower than x). A graphic representation of this idea is shown in figure 1.8. $f(n\nu)$ is the height of the rectangle located at abscissa $n\nu$ while its width is equal to ν, so the area of the nth rectangle separating the curve from the abscissa axis is equal to $f(n\nu)\nu$. When the width of the rectangles tends to zero, the sum of the algebric areas of the rectangles tends to the value of the integral of the function over the considered interval.

A common technique to calculate integrals is called integration by part, and the formula to apply is the following:

$$\int_a^b u'(x)v(x)\,\mathrm{d}x = [u(x)v(x)]_a^b - \int_a^b u(x)v'(x)\,\mathrm{d}x$$

where u' is the derivative of u and v' is the derivative of v.

Another useful property of integrals is their additivity: let a, b, $c \in \mathbb{R}$, we have that:

$$\int_a^c f(x)\,\mathrm{d}x = \int_a^b f(x)\,\mathrm{d}x + \int_b^c f(x)\,\mathrm{d}x \qquad (1.8)$$

1.3.2.2 Use example: averaging

The average value of a function over a given interval is defined as:

$$\langle f(x) \rangle_{[a,\,b]} = \frac{1}{b-a} \int_a^b f(x)\,\mathrm{d}x.$$

1.4 2D functions

1.4.1 Definition

So far we have considered functions that depend on only one parameter. It is in reality very rare that a variable depends only on one parameter! Even something that only varies in space already needs three variables. For that matter, it is of great interest to introduce functions that depend on several parameters. We will limit ourselves here to 2D functions as optics is often observed on a screen, or on our pupil, which are 2D objects. A 2D function is defined as follows:

$$\mathbb{R}, \mathbb{R} \longrightarrow \mathbb{R}$$
$$x, y \longmapsto f(x, y)$$

The concepts exposed before can be adapted to 2D functions.

1.4.2 Graphic representation and examples

Two examples of 2D functions are plotted in figure 1.9, as tridimensional surface plots on the left panels ((a) and (c)). In this kind of graphics, three axes are necessary: two for the independent variables x and y, and one for the dependent variable $f(x, y)$. Some perspective is needed to represent a 3D image on a paper or on a screen, which is why graphics software often offer the possibility to choose the view angle of the virtual camera that takes the picture of that representation. Here in figures 1.9(a) and (c), a colour shade was added for the sake of visibility.

Another way to represent a 2D function is to plot a 'contour map', as shown in figures 1.9(b) and (d) for the same functions as (a) and (c), respectively. In contour maps, we define a contour line of the function $f(x, y)$ as a curve connecting points where the function has a particular value. These contour lines are widely used to represent atmospheric pressures in meteorology which are called isobars, or temperature (isotherms), or altitude of the land in topography maps. Note that in the case of a function that does not depend on y (respectively x), contour lines are normal to the x (respectively y) axis. To visualize the 2D functions on this kind of plot, the value of $f(x, y)$ for a contour line can be associated with a colour or a grey shade, which we will mainly use throughout the book. For example, we can assign the black colour to the lower value of the function and the white colour to the upper value. The intermediate values will have a grey shade according to a linear interpolation between the two extremes and the total number of available grey shades that can be specified by a pixel of the image. This number is determined by the 'bit depth' of the image, referring to the number of bits used to represent each pixel's intensity. A larger bit depth results in a larger number of grey shades and thus to a better resolution of the function representation. For example, with a bit depth of 8 bits, every pixel allows one to represent $2^8 = 256$ shades of grey. Generally, the shades range from black (0) to white (255).

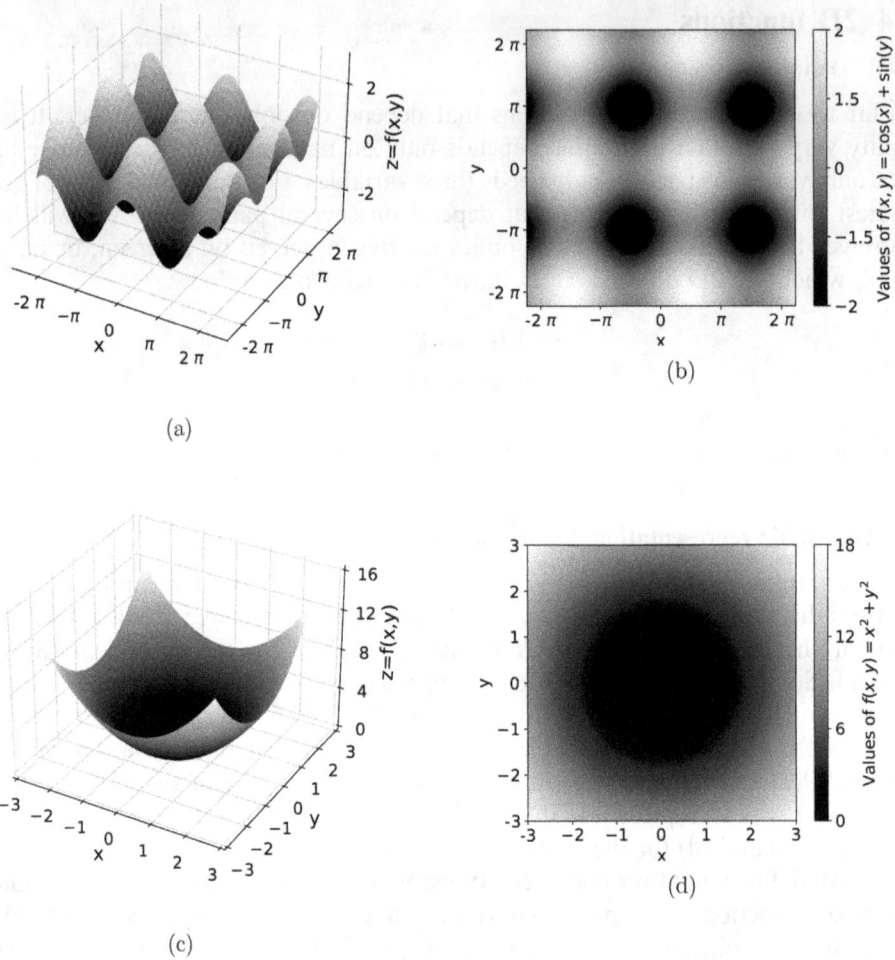

Figure 1.9. Graphic representation of functions depending on two variables x and y as follows: $f(x, y) = \cos(x) + \sin(y)$ as a function of x and y (a) as a 3D surface plot, (b) as a contour plot, and $f(x, y) = x^2 + y^2$ as a function of x and y (c) as a 3D surface plot, (d) as a contour plot.

1.5 Introduction to complex variable functions

1.5.1 Complex numbers

Complex numbers were introduced when we realized that in mathematics, we sometimes needed solutions for equations that require taking the square root of a negative value. In that context, one sets the value i as: $i^2 = -1$.

1.5.1.1 Real and imaginary parts

Complex numbers are defined as:

$$\underline{x} = a + i\,b \text{ where } i^2 = -1$$

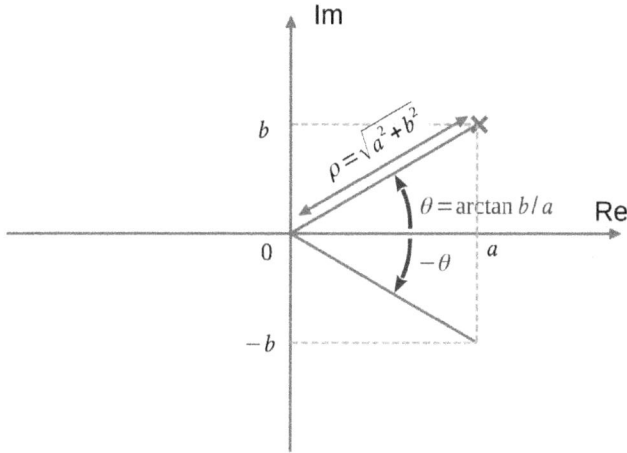

Figure 1.10. Graphic representation of complex numbers.

a is called the real part of the complex number, b is called its imaginary part. That latter definition is called Cartesian, but one can also use the polar (or exponential) representation of a complex number:

$$x = \rho \exp(i\theta) \text{ where } \begin{cases} \rho = \sqrt{a^2 + b^2} \text{ is called the module} \\ \tan \theta = b/a \end{cases}$$

The graphic representation of x in the complex plane is shown in figure 1.10. The ensemble of complex numbers is written $\mathbb{C}$.

1.5.1.2 Conjugate

Every complex number $x = a + ib = \rho \exp(i\theta)$ has a conjugate, defined as $x^* = a - ib = \rho \exp(-i\theta)$. Their real parts are equal while their imaginary parts are opposite. The conjugate of a given complex number is also shown in figure 1.10.

1.5.1.3 Modulus

The module of a complex number is the length of the segment that links the origin of the complex plane to the point of coordinates (a, b). As such, it is equal to $\sqrt{a^2 + b^2}$ (or ρ in the polar definition) and noted $|x|$. Note that two complex numbers that are conjugate have the same module. An interesting property of the module is that for a given complex number $x \cdot x^* = |x|^2$.

1.5.2 Euler relations

The cosine and sine functions, defined earlier, can be expressed as a linear combination of complex numbers. These formulas are called Euler relations:

$$\cos \theta = \frac{e^{i\theta} + e^{-i\theta}}{2} \tag{1.9}$$

$$\sin \theta = \frac{e^{i\theta} - e^{-i\theta}}{2i} \tag{1.10}$$

By recombining these equations we can retrieve another Euler's equation:

$$\exp(i\theta) = \cos\theta + i\sin\theta \tag{1.11}$$

In that context, it is interesting to know that the derivative of an exponential function is the following:

$$\frac{d \exp(\alpha x)}{dx} = \alpha \exp(\alpha x) \text{ where } \alpha \in \mathbb{C}$$

1.5.3 Complex variable functions

A complex variable function f associates a complex number $w = u + iv$ as an image with another complex number $z = x + iy$, as follows:

$$\mathbb{C} \longrightarrow \mathbb{C}$$
$$z \longmapsto w = f(z)$$

where $\mathbb{C}$ refers to the set of all complex numbers.

1.6 Harmonic functions

Mathematically speaking, the complex variable function f defined earlier is said to be harmonic if both its real and imaginary parts respect the Laplace equation:

$$\Delta f = 0 \tag{1.12}$$

where the symbol Δ refers to the Laplace operator. In Cartesian coordinates, equation (1.12) is written:

$$\frac{\partial^2 f}{\partial x^2} + \frac{\partial^2 f}{\partial y^2} + \frac{\partial^2 f}{\partial z^2} = 0$$

A wide variety of mathematical functions respect this equation: for example constants, linear or affine functions, a lot of polynomial functions, etc.

In many domains of physics such as mechanics, electromagnetics, or quantum physics, quantities evolve sinusoidally. In that case, the said quantity can be described by the following expression:

$$g(\rho, \theta) = \rho \cos \theta$$

The argument of the cosine was chosen such that $g(\theta = 2\pi p) = \rho$ where $p \in \mathbb{Z}$ and $g(\theta = 2\pi(p + 1/2)) = -\rho$ where $p \in \mathbb{Z}$.

In that case, one needs to know the expression of (1.12) in polar coordinates, which is:

$$\Delta f(\rho, \theta) = \frac{\partial^2 f}{\partial \rho^2} + \frac{1}{\rho}\frac{\partial f}{\partial \rho} + \frac{1}{\rho^2}\frac{\partial^2 f}{\partial \theta^2} = 0$$

Let us apply the Laplacian in polar coordinates to the latter function:

$$\frac{\partial^2 g}{\partial \rho^2} = \frac{\partial}{\partial \rho}(\cos \theta) = 0$$

$$\frac{1}{\rho}\frac{\partial g}{\partial \rho} = \frac{1}{\rho}\cos \theta$$

$$\frac{1}{\rho^2}\frac{\partial^2 g}{\partial \theta^2} = \frac{1}{\rho^2}\frac{\partial}{\partial \theta}(-\rho \sin \theta) = -\frac{1}{\rho}\cos \theta$$

which leads to $\Delta g(\rho, \theta) = 0$. We conclude that $g(\rho, \theta)$ is a harmonic function.

To prepare for later chapters, where complex exponential notation will simplify the calculations, let us first verify that complex exponential functions are also harmonic functions. In that case, one can write:

$$g(\rho, \theta) = \rho \exp(i\theta) = \rho \cos \theta + \rho i \sin \theta$$

We just showed that the real part of the function, namely $g(\rho, \theta) = \text{Re}(g(\rho, \theta))$, is harmonic. What about the imaginary part $\text{Im}(g(\rho, \theta))$?

$$\frac{\partial^2}{\partial \rho^2}(\rho \sin \theta) = \frac{\partial}{\partial \rho}(\sin \theta) = 0$$

$$\frac{1}{\rho}\frac{\partial}{\partial \rho}(\rho \sin \theta) = \frac{1}{\rho}\sin \theta$$

$$\frac{1}{\rho^2}\frac{\partial^2}{\partial \theta^2}(\rho \sin \theta) = \frac{1}{\rho^2}\frac{\partial}{\partial \theta}(\rho \cos \theta) = -\frac{1}{\rho}\sin \theta$$

from which we conclude that $\text{Im}(g(\rho, \theta))$ is also a harmonic function, and thus the total complex function $g(\rho, \theta)$ is harmonic as well. This important property makes the cosine, sine and complex exponential functions elementary bricks of Fourier analysis, as we will see later in the book.

1.7 Special functions

1.7.1 The 'sinc' function

Let us now define the 'sinc' function:

$$\text{sinc } \theta = \frac{\sin(\theta)}{\theta} \tag{1.13}$$

Its graphical representation is given in figure 1.11. Its maximum, which is reached for $\theta = 0$, is equal to 1 since $\sin(\theta) \underset{\theta \ll 1}{\simeq} \theta$, which divided by θ gives 1. It is an even function that slowly decays to zero while oscillating as $|\theta|$ increases. It will be shown later that this function corresponds to the Fourier transform of a rectangle function, which makes it appear very often in signal processing and in optics.

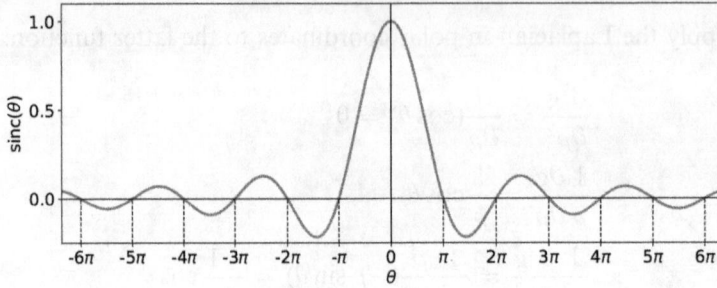

Figure 1.11. Graphic representation of the sinc function.

1.7.2 Bessel functions

Bessel functions are named after the German mathematician Friedrich Bessel. They are solutions to the following differential equation (Bessel's equation [1]), frequently encountered in problems showing a cylindrical symmetry:

$$x^2 \frac{d^2 y_m}{dx^2} + x \frac{dy_m}{dx} + (x^2 - m^2)y_m = 0, \ x \geqslant 0 \text{ and } m \in \mathbb{C}$$

The solutions of this equation are called the Bessel functions. Their expression is:

$$J_m(x) = \frac{1}{\pi} \int_0^\pi \cos(m\phi - x \sin \phi) \, d\phi - \frac{\sin m\pi}{\pi} \int_0^\infty \exp(-x \sinh \theta - m\theta) \, d\theta$$

Let us restrict ourselves to the case where m is an integer. Then $\sin m\pi$ is always zero, and that expression simplifies to:

$$J_m(x) = \frac{1}{\pi} \int_0^\pi \cos(m\phi - x \sin \phi) \, d\phi \qquad (1.14)$$

A graphic representation of the Bessel functions for $m = 1, 2, 3$ is shown in figure 1.12. Note that when m is even, the function is even, while when m is odd, the function is odd as well.

It can be shown that these functions respect the following recurrence relationship [1]:

$$\frac{d}{dx}(x^m J_m(x)) = x^m J_{m-1}(x) \qquad (1.15)$$

Applied to the case where $m = 1$, this becomes:

$$\frac{d}{dx}(x J_1(x)) = x J_0(x)$$

Let's integrate both sides of that last formula between 0 and a:

$$\int_0^a \frac{d}{dx}(x J_1(x)) \, dx = [x J_1(x)]_0^a = a J_1(a) = \int_0^a x J_0(x) \, dx \qquad (1.16)$$

This property will be used later.

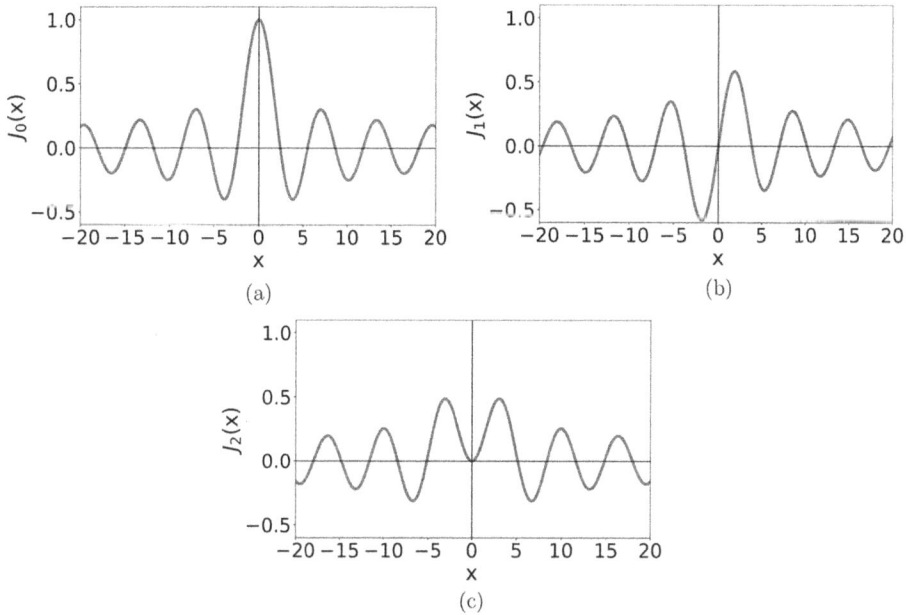

Figure 1.12. Graphic representation of Bessel functions of the first kind J_m for integer orders (a) 0, (b) 1 and (c) 2.

1.7.3 Gaussian function

If we were to plot the intensity of light in a slice of laser beam perpendicular to its propagation direction, we would get what we call a Gaussian function. This mathematical function shows interesting properties and is therefore of high interest in Fourier optics. It owes its name to the German mathematician Johann Carl Friedrich Gauss, who studied probabilities, among many other subjects, where this bell-shaped function is omnipresent. Its mathematical expression is the following:

$$f(x) = \frac{1}{\sigma\sqrt{2\pi}} \exp\left(-\frac{(x - \mu)^2}{2\sigma^2}\right)$$

where μ is called the expected value (the function reaches its maximum at abscissa $x = \mu$) and σ^2 is called the variance. This latter parameter rules the bell's width.

The Gauss function's graphical representations for different sets of parameters are shown in figure 1.13.

1.8 Problems

Problem 1.1: Average of cos, sin, cos² and sin²
Show that the average of a sinusoidal function such as sin or cos over an interval that is much bigger than 2π is equal to zero, while the average of $\sin^2$ or $\cos^2$ over an interval that is much bigger than 2π is equal to 1/2.

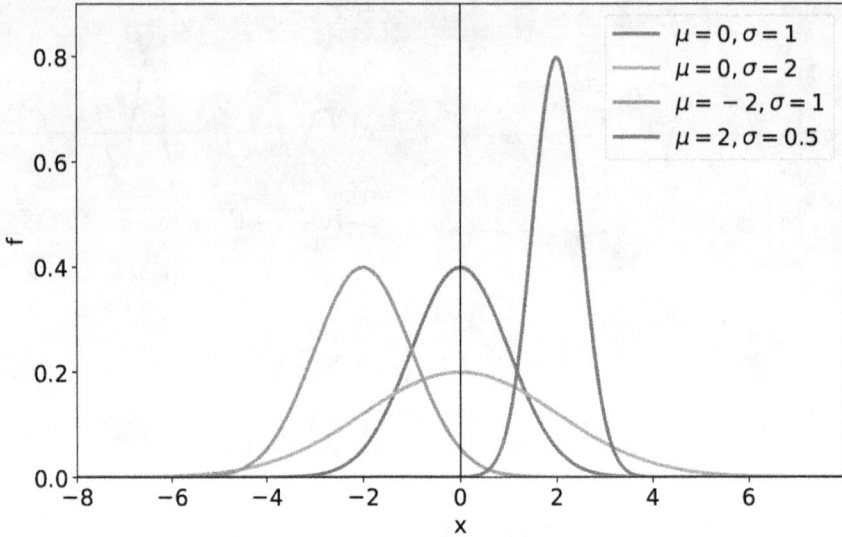

Figure 1.13. Graphic representation of the Gaussian function for different sets of parameters

Solution

$$\frac{1}{b-a} \int_a^b \cos x \, \mathrm{d}x = \frac{1}{b-a} [\sin(x)]_a^b = \frac{1}{b-a}(\sin b - \sin a)$$

and

$$\lim_{(b-a) \to \infty} \left(\frac{1}{b-a}(\sin b - \sin a) \right) = 0$$

A similar calculation shows that the average of sin over a very large interval is zero as well.

Let us now calculate $\frac{1}{b-a} \int_a^b \cos^2 x \, \mathrm{d}x$. We can integrate by part by setting:

$$\begin{cases} u = \cos x \\ v = \sin x \end{cases} \text{ and } \begin{cases} u' = -\sin x \\ v' = \cos x \end{cases}$$

Thus:

$$\frac{1}{b-a} \int_a^b \cos^2 x \, \mathrm{d}x = \frac{1}{b-a} \left([\cos x \sin x]_a^b - \int_a^b - \sin^2 x \, \mathrm{d}x \right)$$

$$= \frac{1}{b-a} \left(\cos b \sin b - \cos a \sin a + \int_a^b \underbrace{\sin^2 x}_{=1-\cos^2 x} \, \mathrm{d}x \right)$$

We recognize the term we are trying to compute on the right: $\frac{1}{b-a} \int_a^b \cos^2 x \, \mathrm{d}x$ and move it to the left side of the equation:

$$2 \times \frac{1}{b-a} \int_a^b \cos^2 x \, dx = \frac{1}{b-a} \left(\cos b \sin b - \cos a \sin a + \underbrace{\int_a^b 1 \, dx}_{=[x]_a^b = b-a} \right)$$

$$= \frac{1}{b-a} (\cos b \sin b - \cos a \sin a) + 1$$

Finally we get:

$$\frac{1}{b-a} \int_a^b \cos^2 x \, dx = \frac{1}{2} \left(\frac{1}{b-a} (\cos b \sin b - \cos a \sin a) + 1 \right)$$

which tends to 1/2 when $b - a$ tends to infinity.

Problem 1.2: Bessel functions expressions
Using Euler's formulas, parity properties of cosine and sine functions, and some calculus rules for integrals, demonstrate that formula (1.14) is equal to:

$$J_m(x) = \frac{i^{-m}}{2\pi} \int_0^{2\pi} \exp(i(mv + x \cos v)) \, dv \qquad (1.17)$$

Solution:

$$J_m(x) = \frac{i^{-m}}{2\pi} \int_0^{2\pi} \exp(i(mv + x \cos v)) \, dv$$

$$= \frac{\exp\left(-i\frac{\pi}{2}m\right)}{2\pi} \int_0^{2\pi} \exp(i(mv + x \cos v)) \, dv$$

$$= \frac{1}{2\pi} \int_0^{2\pi} \exp(i(mv + x \cos v - m\pi/2)) \, dv$$

We then implement the following change of variable: $v \leftarrow v - \pi/2$:

$$J_m(x) = \frac{1}{2\pi} \int_{-\pi/2}^{3\pi/2} \exp(i(mv + x \cos(v + \pi/2))) \, dv$$

We have that $\cos(v + \pi/2) = -\sin v$, leading to:

$$J_m(x) = \frac{1}{2\pi} \int_{-\pi/2}^{3\pi/2} \exp(i(mv - x \sin v)) \, dv$$

$$= \frac{1}{2\pi} \left[\int_{-\pi/2}^{3\pi/2} \cos(mv - x \sin v) \, dv + i \int_{-\pi/2}^{3\pi/2} \sin(mv - x \sin v) \, dv \right]$$

(from Euler's formula, equation (1.11)).
Since $\cos(mv - x \sin v)$ is identical over $[\pi, 3\pi/2]$ as over $[-\pi, -\pi/2]$, and since $\sin(mv - x \sin v)$ is also identical over $[\pi, 3\pi/2]$ as over $[-\pi, -\pi/2]$, we can replace the integrals limits as follows:

$$J_m(x) = \frac{1}{2\pi}\left[\underbrace{\int_{-\pi}^{\pi}\cos(mv - x\sin v)\,dv}_{A} + i\underbrace{\int_{-\pi}^{\pi}\sin(mv - x\sin v)\,dv}_{B}\right]$$

We can use equation (1.8) for integrals to compute A:

$$A = \int_{-\pi}^{0}\cos(mv - x\sin v)\,dv + \int_{0}^{\pi}\cos(mv - x\sin v)\,dv$$

We then implement the following change of variable: $v \leftarrow -v$ on the left term of that last integral:

$$\int_{-\pi}^{0}\cos(mv - x\sin v)\,dv = -\int_{\pi}^{0}\cos(-mv + x\sin v)\,dv \text{ since } \sin{-u} = -\sin u$$

$$= \int_{0}^{\pi}\cos(mv - x\sin v)\,dv \text{ since } \cos{-u} = \cos u$$

Finally:

$$A = 2\int_{0}^{\pi}\cos(mv - x\sin v)\,dv$$

Following similar steps, we get that $B = 0$. To conclude, we obtain:

$$J_m(x) = \frac{1}{\pi}\int_{0}^{\pi}\cos(mv - x\sin v)\,dv$$

Reference

[1] Selcuk Bayin S 2008 *Essentials of Mathematical Methods in Science and Engineering* (New York: Wiley)

IOP Publishing

Elementary Fourier Optics for Science and Engineering Students

Hélène Ollivier and Osvaldo de Melo

Chapter 2

Optics fundamentals—a review

In this chapter, we introduce the basics in optics, laying the foundation for the subsequent chapters. We begin with the fundamental concepts of ray optics and their applications in optical systems. Then, we define waves and present their mathematical and graphical representations, along with their key parameters: amplitude, frequency and wavelength. We apply this formalism to light, which is an electromagnetic wave. Using the complex notation from chapter 1, we derive light intensity from the electromagnetic field. This naturally leads to interference under some conditions that we establish. We illustrate these principles with Young's double-slit experiment. Next, we introduce diffraction, a fundamental concept of Fourier optics. We discuss Huygens' and Fresnel's contributions to the formalism of diffraction, and then present Fresnel and Fraunhofer approximations and their assumptions. Finally, we derive the behaviour of light diffracted by various apertures: single slit, double-slit, rectangular, and circular.

2.1 Basic concepts of geometric optics

Optics has been fascinating since humans were aware that they could see. The field itself started developing as early as about four centuries BC, when Euclid introduced the idea of 'optical ray'. Around this time was raised a question: some intellectuals thought that their eyes contained a 'pure fire' that allowed them to send light rays to objects, and others, including Plato, were convinced that enlightened objects were sending light particles to our eyes. Later on was developed the model of geometric optics, according to which light travels in a vacuum and transparent media, following a linear path.

2.1.1 Speed of light

In a vacuum, the speed of light is precisely defined as 299 792 458 metres per second, a fundamental constant in physics. The metre is derived from this value, based on the defined duration of a second, which corresponds to 9 192 631 770 oscillations

doi:10.1088/978-0-7503-6392-1ch2
2-1

between two energy states of a caesium 133 atom [1]. For every transparent material, namely every material that light radiation can pass through, the refractive index n is defined as the ratio of the speed of light v in that material, to the speed of light c in a vacuum:

$$n = \frac{v}{c}$$

2.1.2 Snell's laws

When a ray of light encounters an interface, it gives rise to two rays: one of them is reflected while the other one is transmitted through the interface, and deviates if the two media separated by the interface do not have the same refractive index. In that latter case, the ray is said to be refracted. The behaviours of the reflected and refracted rays are ruled by the three Snell's laws of reflection and refraction. The first law states that the reflected ray and the refracted ray both belong to the plane defined by the normal to the interface and the incident ray (see figure 2.1).

2.1.2.1 Reflection
The second law gives the reflection angle between the reflected ray and the normal to the interface. Let θ_i be the incidence angle, then the reflection angle is equal to $\theta_r = -\theta_i$ (see figure 2.1).

2.1.2.2 Refraction
The third law gives the refraction angle between the refracted ray and the normal to the interface. The refraction angle is given by $n_1 \sin \theta_i = n_2 \sin \theta_t$ (see figure 2.1).

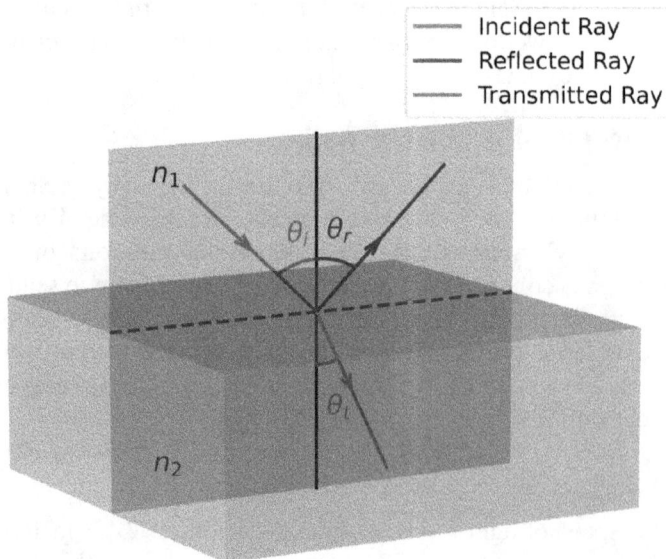

Figure 2.1. Representation of Snell's laws.

2.2 Lenses and image formation

2.2.1 Converging lenses

The refraction phenomenon allows us to shape the light: let us consider a piece of glass that is flat on one side for the sake of simplicity, and spherical on the other side, as shown in figure 2.2. An incident light beam entering the glass perpendicularly to the flat surface would not undergo any refraction. However, at the second interface, when the light comes out of the glass on the spherical side, it undergoes a refraction whose angle is not the same from one ray to another (it depends on the vertical coordinate in figure 2.2). By applying Snell's third law to each of the incoming beams, one can compute the trajectory of the emerging light.

We can see that the rays do not meet on a single point. This phenomenon, called spherical aberration, tends to disappear as the lens is very thin (see figure 2.3). In that case, the light turns out to be focussing in a point, called the focal point.

Several solutions exist to minimize spherical aberrations [2]. For example, using bi-convex lenses instead of lenses that are flat on one side and convex on the other one, as we just considered, decreases the incidence angles of light on the surfaces and thus decreases the spherical aberration. Using several lenses instead of just one also has the same effect, although it shows the disadvantage of increasing the cost, the space occupation and the weight of the optical system. Another solution would be to use lenses made of a higher index material, which would allow a smaller radius of curvature for a given focal length. Finally, it is possible (although much more

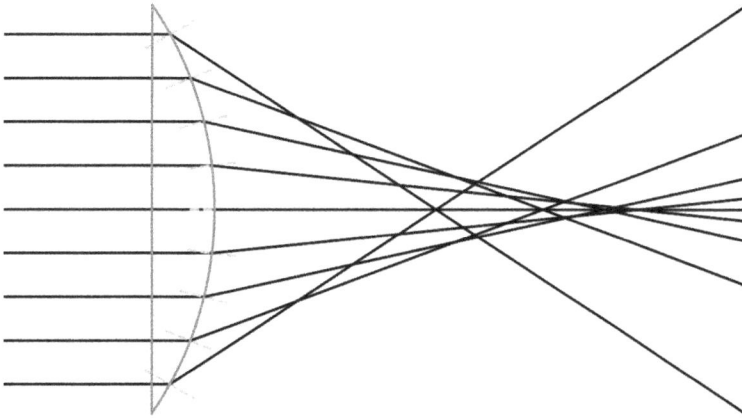

Figure 2.2. Refraction in a converging half-spherical lens.

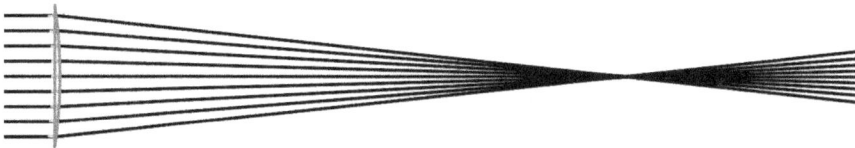

Figure 2.3. Focalization of an incoming beam of rays by a thin converging half-spherical lens.

complicated) to make the lenses' surfaces aspherical, by grinding, polishing, or moulding the material. The curvature can then be engineered as a function of surface position, to erase any spherical aberration.

2.2.2 Diverging lenses

Just as a convex lens can be used to converge light, a concave lens can be used to make it diverge, as shown in figure 2.4.

If we plot the rays upstream until the point that seems to be their source, we can see that here again they do not meet at a single point (see figure 2.5): spherical aberration also occurs with diverging spherical lenses. As well as in the converging lens case, it tends to disappear as the lens is very thin (see figure 2.6).

2.2.3 Lenses in optical systems

We saw in the last section that when lenses are thin enough, they allow one to focus a collimated beam of light (made of rays that are all parallel to each other) towards a

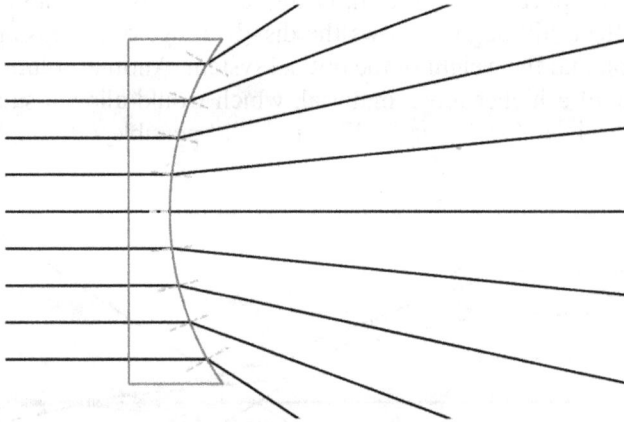

Figure 2.4. Refraction in a diverging half-spherical lens.

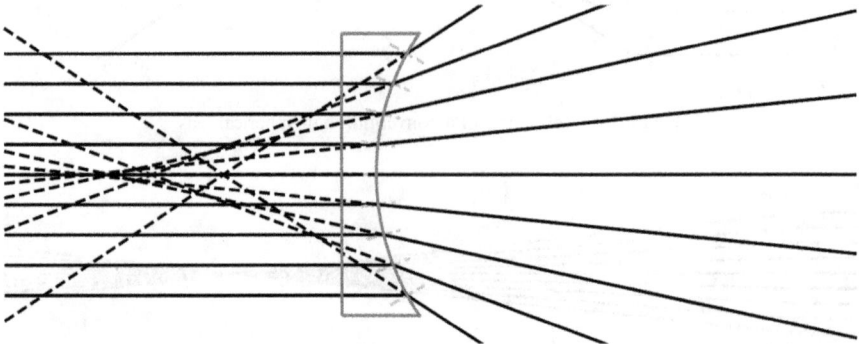

Figure 2.5. Refraction in a diverging half-spherical lens, with extension of the beams upstream.

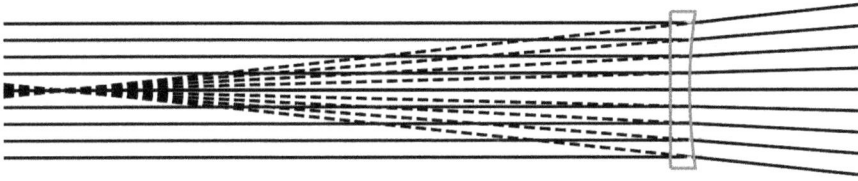

Figure 2.6. Focalization of an incoming beam of rays by a thin diverging lens.

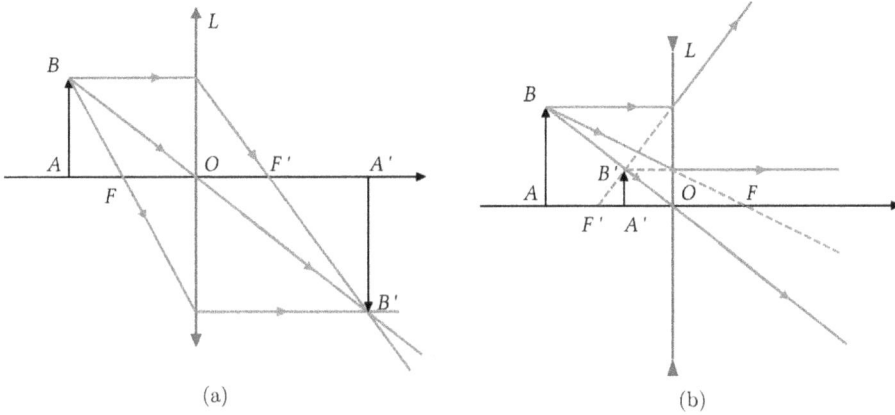

(a)   (b)

Figure 2.7. Representation of the optical rays coming from an object and crossing (a) a thin converging lens and (b) a thin diverging lens.

discrete point, called the image focal point. When light coming from an object AB perpendicular to the optical axis, where A is on the optical axis and B is close enough to the optical axis reaches a thin lens, the latter gives an image $A'B'$ whose position is related to the object position through what is called the thin-lens equation:

$$\frac{1}{\overline{OA'}} - \frac{1}{\overline{OA}} = \frac{1}{\overline{OF'}}$$

where O is the centre of the thin lens, A refers to the position of the object, A' to the position of the image, and F' to the image focal point. The situation is represented in figure 2.7(a) for the case of a converging lens, where the usual rules of plotting optical rays are shown: the ray that comes from B and crosses the thin lens through its centre is not deviated, while the one that comes from B and arrives on the lens parallel to the optical axis crosses the image focal point. The intersection of these two resulting beams shows the position of B'.

As we can see in figures 2.7(a) and (b), the size of the image and the size of the object are related through the following relationship:

$$\frac{\overline{A'B'}}{\overline{AB}} = \frac{\overline{OA'}}{\overline{OA}} = \gamma, \text{ called the lens magnification} \qquad (2.1)$$

The case of thin diverging lenses is very similar: the difference lies in the fact that the image focal point is on the same side as the object. The thin-lens relationship is still valid, as well as equation (2.1). The rules to draw the optical rays remain the same. An example is shown in figure 2.7(b). In the latter case, we can see that the image is formed before the lens. In that case, we could not see that image on a screen and the image is called a virtual image. Although it cannot be observed on a screen, it can be used as an object for another lens!

2.3 Basic concepts of waves

2.3.1 Definition

A wave is a periodic phenomenon that carries energy without carrying matter. We can mathematically refer to it as a coupling between space and time, since the carried quantity varies with space (if we take a picture at a given time of the wave, we see a periodicity) and with time (if we post ourselves at a point of space where the wave passes, we observe a variation in time of the carried quantity). A trivial example of waves is ocean waves: they consist of a height of sea water level that couples time and space. If we take a picture of the ocean during a storm, we see a periodic evolution of the sea level in space in a given direction, and if we stay still in the sea, we move up and down without moving in the two other directions of space. There exist many types of waves: sound waves, that consist of periodic compression of material slices, electromagnetic waves (such as light [3]) that are a periodic electromagnetic field, gravitational waves, that are the propagation of distortions of space-time [4], probability waves in quantum mechanics, and there are probably more to discover.

2.3.2 Mathematical expression: amplitude, frequency, wavelength

To illustrate the basic concepts of waves, we will consider the case of a plane wave. In that case, the coupling between space ($\vec{r}$) and time (t) for a given quantity $Q(\vec{r}, t)$ mathematically translates as the following expression for the considered quantity (height of the wave at a given point in space $\vec{r}$ and time t):

$$Q(\vec{r}, t) = Q_0 \cos(\Phi(\vec{r}, t))$$
$$= Q_0 \cos(\vec{k} \cdot \vec{r} - \omega t + \phi) \tag{2.2}$$

$Q(\vec{r}, t)$ is called the wave amplitude. Q_0 is called the maximum amplitude, as it refers to the maximal value that $Q(\vec{r}, t)$ can take. The argument of cosine, $\Phi(\vec{r}, t)$, is referred to as the phase of the wave. It is equal to $\vec{k} \cdot \vec{r} - \omega t + \phi$, where the different components are defined below:

• The vector $\vec{k}$ is called the wave vector. Its direction is that of the propagation, and its modulus is equal to $k = 2\pi/\lambda$ where λ is the spatial period of the wave, called the wavelength. The inverse of λ is called the spatial frequency.

• ω is called the wave angular frequency and is related to its frequency f by $\omega = 2\pi f$.

• Finally, ϕ is an additional parameter that allows one to take into account the possibly existing phase at $\vec{r} = 0$, $t = 0$ of the wave quantity value.

Note that equation (2.2) has the form of a harmonic function (see section 1.6). It is common to call this specific type of wave a harmonic plane progressive wave.

2.3.3 Graphical representation

In expression (2.2), let us detail the argument of the cosine:

$$\vec{k} \cdot \vec{r} - \omega t + \phi = k_x x + k_y y + k_z z - \omega t + \phi$$

For the sake of simplicity, let us consider that the wave propagates along the z direction only. In that case, the wave vector components along the x and y directions are zero ($k_x = k_y = 0$) and $k_z = k$. Let us further assume that the phase of the wave at $\vec{r} = 0$, $t = 0$ is zero, namely $\phi = 0$. In that case, the expression of the wave amplitude reduces to:

$$Q(z) = Q_0 \cos(kz - \omega t)$$

The graphic representation of a such a progressing wave is shown in figure 2.8.

2.3.4 Phase velocity

The phase velocity is the speed at which a fixed phase point of the wave moves through space. To determine this velocity, we consider a constant phase in expression (2.2). For $kz - \omega t$ to remain constant as time evolves, the position z must satisfy $z = \omega t / k$. Therefore, the phase velocity is mathematically defined as:

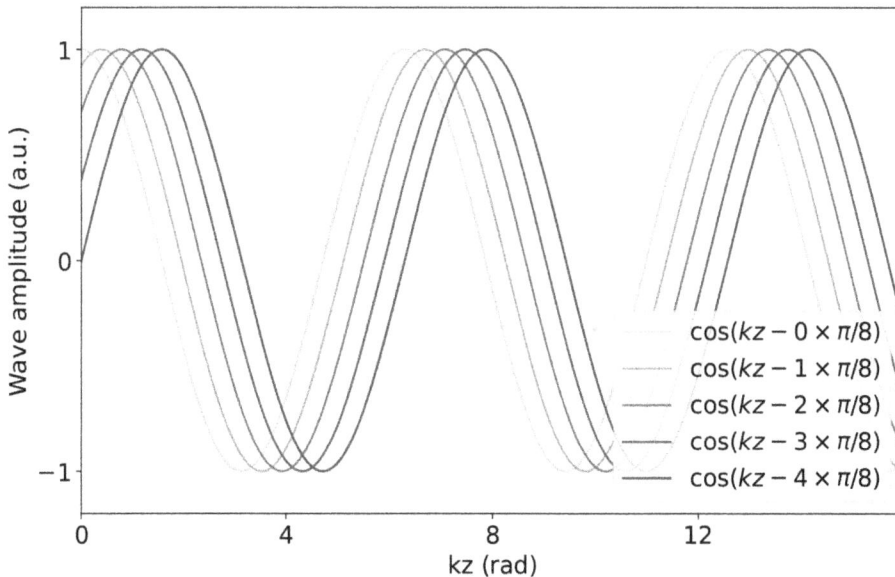

Figure 2.8. Graphical representation of the wave progression: its amplitude is plotted as a function of space (like a picture) as time evolves (from light to dark blue). Here the original phase ϕ was chosen to be zero.

$$v_\Phi = \frac{\omega}{k}$$

Note that using the previously given expressions $\omega = 2\pi f$ and $k = 2\pi/\lambda$ leads to $v_\Phi = \lambda f$ and for an electromagnetic field, the wavelength inside a material of refractive index n is equal to λ_0/n where λ_0 is the wavelength of that electromagnetic field in a vacuum. Finally, that gives us $v_\Phi = c/n$ where $c = \lambda_0 f$ is the speed of light in a vacuum.

2.3.5 Wavefronts and Malus's law

The wavefronts are defined as the surfaces formed by all the points of equal phase. For example, the wavefronts for a plane wave are planes, and the wavefronts for a spherical wave are spheres. These two examples are shown in figures 2.9 and 2.10.

Malus's law states that the light rays are orthogonal to the wavefronts in every point of space.

2.3.6 Exponential representation

There exists a powerful tool to simplify calculation in wave optics: the exponential representation $\underline{Q}(\vec{r}, t)$ of a wave quantity $Q(\vec{r}, t)$. Using Euler's formula (see section 1.5.2), we can write the wave amplitude at position $\vec{r}$ and time t as follows:

$$\begin{aligned} Q(\vec{r}, t) &= \frac{Q_0}{2}(\exp(i\Phi(\vec{r}, t)) + \exp(-i\Phi(\vec{r}, t))) \\ &= \frac{Q_0}{2}(\exp(i(\vec{k}\cdot\vec{r} - \omega t + \phi)) + \exp(-i(\vec{k}\cdot\vec{r} - \omega t + \phi))) \end{aligned}$$

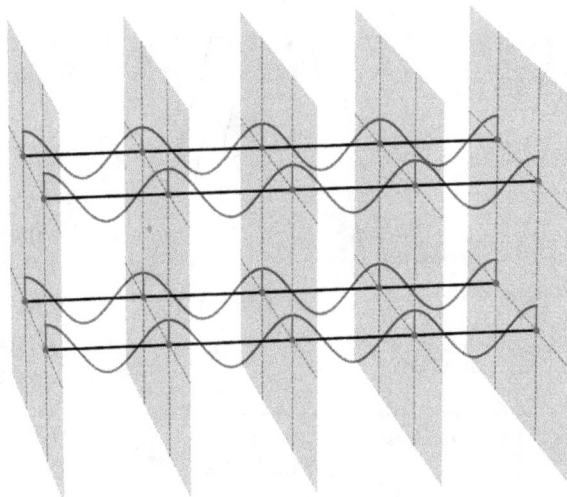

Figure 2.9. Plane wavefronts for a collimated source (emitting a plane wave).

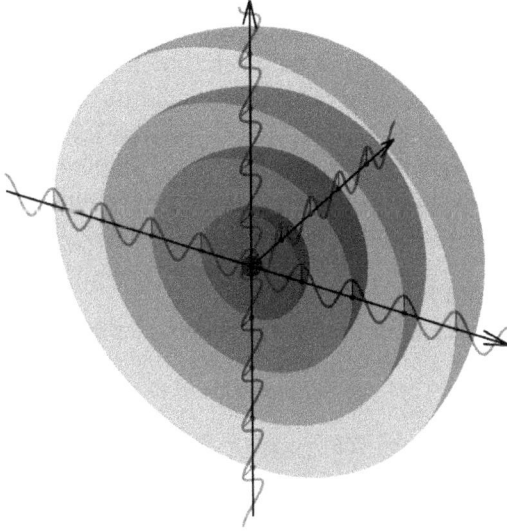

Figure 2.10. Hemispherical wavefronts for a punctual source emitting in half of the space directions.

which we often write as follows:

$$Q(\vec{r}, t) = \frac{1}{2}(\underline{Q}(\vec{r}, t) + \underline{Q}^*(\vec{r}, t))$$

where

$$\underline{Q}(\vec{r}, t) = Q_0 \exp{(i\Phi(\vec{r}, t))} = Q_0 \exp{(i(\vec{k} \cdot \vec{r} - \omega t + \phi))}$$

$\underline{Q}(\vec{r}, t)$ is called the complex amplitude of the field.

Using this expression instead of the cosine expression facilitates the calculations since it does not require the use of trigonometric formulas. One can also write the latter formula as:

$$Q(\vec{r}, t) = \frac{1}{2}(\underline{Q}(\vec{r}, t) + \text{ c.c. })$$

where c.c. refers to complex conjugate.

2.4 Interference of light

2.4.1 Electric field amplitude and light intensity

Light is an electromagnetic wave, meaning it consists of oscillating electric and magnetic fields that propagate through space. The electric field component of this wave can be expressed as:

$$\vec{E}(\vec{r}, t) = \vec{E_0} \cos(\vec{k} \cdot \vec{r} - \omega t + \phi)$$

Our usual detectors (our eyes, photodiodes, photoresistances) are not sensitive to the amplitude of optical waves but to its intensity. Light intensity is expressed as follows:

$$I = K \langle ||\vec{E}(\vec{r}, t)||^2 \rangle_{\tau_D}$$

where the real expression of the field was used. We can show that the multiplicative constant K is equal to $\varepsilon_0 c$ where ε_0 is the electric permittivity of vacuum and c is the light speed in vacuum. The double vertical bars refer to the magnitude of the vector, and the angle brackets denote an average (see section 1.3.2.2), here taken over the detector response time τ_D. The unit of I is W m^{-2}. Note that the term intensity in radiometry refers to the light flux per unit of solid angle (in W sr^{-1}), while the irradiance is the light flux per unit of area (W m^{-2}). In this book we will use the term intensity as we are not focussing on radiometry notions, but it should be known that radiometry people would be a bit mad at us.

From there, let us determine the light intensity at position $\vec{r}$ and time t:

$$
\begin{aligned}
I(\vec{r}, t) &= K \langle \| \vec{E_0} \cos(\vec{k} \cdot \vec{r} - \omega t + \phi) \|^2 \rangle_{\tau_D} \\
&= K E_0^2 \left\langle \cos^2(\vec{k} \cdot \vec{r} - \omega t + \phi) \right\rangle_{\tau_D}
\end{aligned}
$$

The temporal period of the wave T, which is of the order of 10^{-15} s for optical waves, is much smaller than the detection time, e.g. $T \ll \tau_D$ (for the quickest detectors, this value is about 10^{-9} s), then we can state that the average value of the above cosine squared is equal to $1/2$ (see solution of problem 1.1) Then:

$$
I(\vec{r}, t) = \frac{K E_0^2}{2}
$$

In complex notation, we have:

$$
\vec{E}(\vec{r}, t) = \frac{1}{2}(\vec{E_0} \exp(i(\vec{k} \cdot \vec{r} - \omega t)) + \text{c.c.}) = \frac{1}{2}(\underline{\vec{E}} + \underline{\vec{E}}^*)
$$

From which the intensity can also be calculated:

$$
I = K \langle \| \vec{E}(\vec{r}, t) \|^2 \rangle_{\tau_D} = K \left\langle \| \frac{1}{2}(\underline{\vec{E}} + \underline{\vec{E}}^*) \|^2 \right\rangle_{\tau_D} = \frac{K}{4} \langle \| (\underline{\vec{E}} + \underline{\vec{E}}^*) \|^2 \rangle_{\tau_D}
$$

We can then use the following property of complex numbers: $\| \underline{x} \|^2 = \underline{x} \cdot \underline{x}^*$ (see section 1.5.1), which applied here leads to:

$$
\| \underline{\vec{E}} + \underline{\vec{E}}^* \|^2 = (\underline{\vec{E}} + \underline{\vec{E}}^*)(\underline{\vec{E}}^* + \underline{\vec{E}}) = (\underline{\vec{E}} + \underline{\vec{E}}^*)^2 = \underline{\vec{E}}^2 + \underline{\vec{E}}^{*2} + 2\underline{\vec{E}} \cdot \underline{\vec{E}}^*
$$

Using the expression of $\underline{\vec{E}} = \vec{E_0} \exp i\Phi$, that leads to:

$$
I = \frac{K}{4} \left\langle \underbrace{E_0^2 \exp(2i\Phi) + E_0^2 \exp(-2i\Phi)}_{=2E_0^2 \cos(2\Phi)} + 2E_0^2 \underbrace{\exp(i(\Phi - \Phi))}_{=1} \right\rangle_{\tau_D}
$$

The term $2E_0^2 \cos(2\Phi)$ oscillates in time with an angular frequency equal to 2ω, which means it vanishes when the intensity is averaged over the response time of the detector, as mentioned before. We thus retrieve the same result as when using the real notation:

$$I = \frac{KE_0^2}{2}$$

Actually, when working with optical waves, we are always under the assumption that the sum of any combination of optical angular frequencies is much bigger than $2\pi/\tau_D$ (the detector bandwidth), as we assumed in the two previous cases. It implies that the expression for the intensity can always, within the concepts treated in this textbook, be taken as:

$$I = \frac{K}{2}\vec{E} \cdot \vec{E}^* \tag{2.3}$$

where the temporal averaging over the time response of the detector has been dropped, considering that terms oscillating with a temporal frequency equal to a difference of two optical frequencies vary slowly enough for the detector to follow.

2.4.2 Sum of light waves

When several waves arrive at the same point in space, one has to add up the waves amplitudes before calculating the intensity. The obtained result is not always equal to the sum of the intensities of the waves amplitudes calculated separately beforehand. The difference is called the interference term. To illustrate, let us consider the simple case where two waves of the same angular frequency meet at a position $\vec{r}$. Let their real expressions be:

$$\vec{E_1} = \vec{E}_{01}\cos(\vec{k_1} \cdot \vec{r} - \omega_1 t + \phi_1(t)) = \vec{E}_{01}\cos\Phi_1(\vec{r}, t) \tag{2.4}$$

$$\vec{E_2} = \vec{E}_{02}\cos(\vec{k_2} \cdot \vec{r} - \omega_2 t + \phi_2(t)) = \vec{E}_{02}\cos\Phi_2(\vec{r}, t) \tag{2.5}$$

The intensity at point $\vec{r}$ and time t is:

$$I_{1+2} = K\langle ||\vec{E_1}(\vec{r}, t) + \vec{E_2}(\vec{r}, t)||^2 \rangle_{\tau_D}$$

The first term arising from the term distribution is calculated as the intensity of a single wave and is equal to $KE_{01}^2/2$. Similarly, the last term gives $KE_{02}^2/2$. The crossed term is equal to:

$$K\vec{E}_{01} \cdot \vec{E}_{02}(\cos(\Phi_1(\vec{r}, t) + \Phi_2(\vec{r}, t)) + \cos(\Phi_1(\vec{r}, t) - \Phi_2(\vec{r}, t)))$$

When averaging the interference term over the detector response time, the term $\cos(\Phi_1(\vec{r}, t) + \Phi_2(\vec{r}, t))$ disappears, as its period $2\pi/(\omega_1 + \omega_2)$ in the case of optical waves is about 10^{-15} s $\ll \tau_D$.

Finally, using a trigonometric relation from section 1.2.3:

$$I_{1+2} = \frac{K}{2}(E_{01}^2 + E_{02}^2 + \underbrace{2\vec{E}_{01} \cdot \vec{E}_{02}\cos(\Phi_1(\vec{r}, t) - \Phi_2(\vec{r}, t))}_{\text{interference term}}) \tag{2.6}$$

Let us show that expression (2.3) and the use of complex exponential notation for the electric fields lead to the same result with a shorter calculation:

$$I_{1+2} = \frac{K}{2}(\vec{E_1}+\vec{E_2})(\vec{E_1}^*+\vec{E_2}^*) \text{ where } \begin{cases} \vec{E_1} = \vec{E_{01}}\exp(i\Phi_1(\vec{r}, t)) \\ \vec{E_2} = \vec{E_{02}}\exp(i\Phi_2(\vec{r}, t)) \end{cases}$$

$$= \frac{K}{2}\left(E_{01}^2 + E_{02}^2 + \vec{E_{01}}\vec{E_{02}} \underbrace{(\exp(-i\Phi_1(\vec{r}, t) + i\Phi_2(\vec{r}, t)) + \exp(i\Phi_1(\vec{r}, t) - i\Phi_2(\vec{r}, t)))}_{2\cos(\Phi_1(\vec{r},t)-\Phi_2(\vec{r},t))} \right)$$

This technique becomes even more powerful in cases where more than two waves need to be taken into account.

2.4.3 Conditions for interference

From expression (2.6) arise three conditions that are necessary to observe interference.

2.4.3.1 Condition on polarization

The electric field is a vector. It has a direction in space, which is called its polarization. The term $\vec{E_{01}} \cdot \vec{E_{02}}$ from expression (2.6) shows that a necessary condition to observe interference is that the two electric fields $\vec{E_1}$ and $\vec{E_2}$ are not orthogonal to each other. In other words, the incoming waves have to not be perpendicularly polarized.

2.4.3.2 Condition on dephasings

As we can see in the expression of the fields (2.4) and (2.5), the parameters ϕ_1 and ϕ_2 actually depend on time. This translates the fact that a light source emits wave trains: every once in a while, a discrete event 'reinitializes' the emission process, making the phases $\phi_1(t)$ and $\phi_2(t)$ randomly dependent on time. If these events happen many times over the time response of the detector, the interference term vanishes. This can be avoided by maintaining $\phi_1(t)$ and $\phi_1(t)$ equal so that $\phi_1(t) - \phi_2(t) = 0$ at all times. This condition on dephasings can be fulfilled by phase-locking two sources or by using one source and dividing its light stream into two, to make them interfere.

2.4.3.3 Condition on angular frequency

Given that the previous condition for interference is fulfilled (namely that the dephasings $\phi_1(t)$ and $\phi_2(t)$ evolve together), the argument of cosine in expression (2.6) is equal to:

$$\Phi_1(\vec{r}, t) - \Phi_2(\vec{r}, t) = (\vec{k_1} - \vec{k_2}) \cdot \vec{r} - (\omega_1 - \omega_2)t + \underbrace{\phi_1(t) - \phi_2(t)}_{=0}$$

$$= (\vec{k_1} - \vec{k_2}) \cdot \vec{r} - (\omega_1 - \omega_2)t$$

Thus, the intensity I_{1+2} oscillates in time with an angular frequency $\omega_1 - \omega_2$, so with a period $T = 2\pi/(\omega_1 - \omega_2)$. If $\omega_1 - \omega_2$ is small enough compared to $2\pi/\tau_D$, we can observe a variation of I_{1+2} in time. Otherwise, the interference term cannot be observed since its average over τ_D is equal to $1/2$. Note that using two sources that have the same angular frequency (or one source where we split the stream of light in two parts and then recombine them) cancels the $(\omega_1 - \omega_2)t$ term and leads to a constant intensity. In that case, we can observe interferences in space since $I_{1+2}(\overrightarrow{r}, t) = I_{1+2}(\overrightarrow{r})$.

2.4.3.4 Coherence

In the case where all the interference conditions are fulfilled, the sources are said to be coherent. If the sources are sufficiently incoherent, it is a good approximation to calculate the total intensity by simply summing the individual intensities of each wave separately. Indeed the interference term in equation (2.6) vanishes and we are left with:

$$I_{1+2} = I_1 + I_2 \text{ where } \begin{cases} I_1 = \dfrac{KE_{01}^2}{2} \\ I_2 = \dfrac{KE_{01}^2}{2} \end{cases}.$$

2.4.4 Phase difference and optical path

To calculate the term $\cos(\Phi_1(\overrightarrow{r}, t) - \Phi_2(\overrightarrow{r}, t))$ one has to determine the phase difference between two beams arriving at the same position $\overrightarrow{r}$ at time t after having followed different paths. To do that, we have to calculate the accumulated phase by each of the beams and take the difference. How do we determine the accumulated phase by one beam that followed a given path? We consider a picture of the situation at $t = t_0$. The phase accumulated by the optical wave on an infinitesimal linear element dx of a given path C that has been travelled by the wave is given by:

$$\Phi(x + dx) - \Phi(x) = (k(x + dx) - \omega t_0 + \phi) - (kx - \omega t_0 + \phi) = k\,dx = \frac{2\pi}{\lambda_0} n(x)\,dx$$

where we assumed that the refractive index n is constant between x and $x + dx$ and equal to $n(x)$. For example, the phase accumulated by a beam following a straight path from a point A at abscissa x_A to a point B at abscissa x_B with no medium change is given by:

$$\Phi_{acc} = \int_{x_A}^{x_B} \frac{2\pi}{\lambda_0} n(x)\,dx = \frac{2\pi}{\lambda_0} n(x_B - x_A)$$

More generally, if the path is curved, we can cut it down into infinitesimally small straight paths (see figure 2.11) and modify the previous integral into what is called a line integral as follows:

$$\Phi_{acc} = \frac{2\pi}{\lambda_0} \underbrace{\int_C n(s)\,ds}_{\text{optical path length } \delta}$$

where s refers to the curvilinear abscissa along the path. The term $\int_C n(s)\,ds$ is called the optical path length. Given that $n = c/v$ and $v = \frac{ds}{dt}$, we have $n\,ds = c\,dt$, which

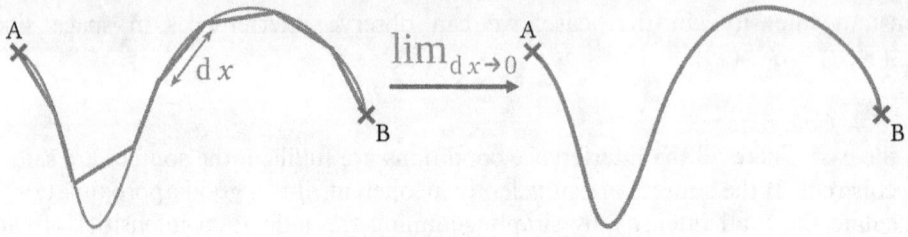

Figure 2.11. Schematization of a curved path as the limit when dx tends towards zero of a series of straight lines, for the calculation of the accumulated phase of a light beam along that curved path.

implies that the optical path length can be seen as the distance the beam would have travelled in the same duration if it were travelling in a vacuum. As a consequence, the phase difference $\Delta\Phi$ between two beams having followed different trajectories is finally equal to $\frac{2\pi}{\lambda_0}\Delta\delta$ where $\Delta\delta$ is the optical path length difference between the two considered light beams.

2.4.5 Constructive and destructive interference

Let us now consider two incoming waves that fulfil all the interference conditions. They come from one only source, which means that they have the same angular frequency ($\omega = \omega_1 = \omega_2$) and phase at the origin ($\phi_1(t) = \phi_2(t)$ for all t). Their polarizations are parallel. In that case, expression (2.6) resumes to:

$$I_{1+2} = \frac{K}{2}\left(E_{01}^2 + E_{02}^2 + 2E_{01}E_{02}\cos\left(\frac{2\pi}{\lambda_0}\Delta\delta\right)\right) \qquad (2.7)$$

The intensity evolves in space. As cos varies from -1 to $+1$, the intensity varies from $I_{\min}$ (we talk about 'destructive interference') to $I_{\max}$ ('constructive interference') with:

$$\begin{cases} I_{\min} = \frac{K}{2}(E_{01}^2 + E_{02}^2 - 2E_{01}E_{02}) = (\sqrt{I_1} - \sqrt{I_2})^2 \\ I_{\max} = \frac{K}{2}(E_{01}^2 + E_{02}^2 + 2E_{01}E_{02}) = (\sqrt{I_1} + \sqrt{I_2})^2 \end{cases} \text{where} \begin{cases} I_1 = \frac{K}{2}E_{01}^2 \\ I_2 = \frac{K}{2}E_{02}^2 \end{cases}$$

I_1 and I_2 represent the intensities of beams 1 and 2 if they were not interfering.

Destructive interference (see right panel of figure 2.12) happens when the cosine is -1, which is the case whenever its argument $\Delta\Phi$ is an odd multiple of π, which corresponds to a path difference equal to:

$$\Delta\Phi = (2p + 1)\pi \text{ where } p \in \mathbb{Z}$$

which corresponds to a path difference equal to:

$$\Delta\delta = \left(p + \frac{1}{2}\right)\lambda_0 \text{ where } p \in \mathbb{Z}$$

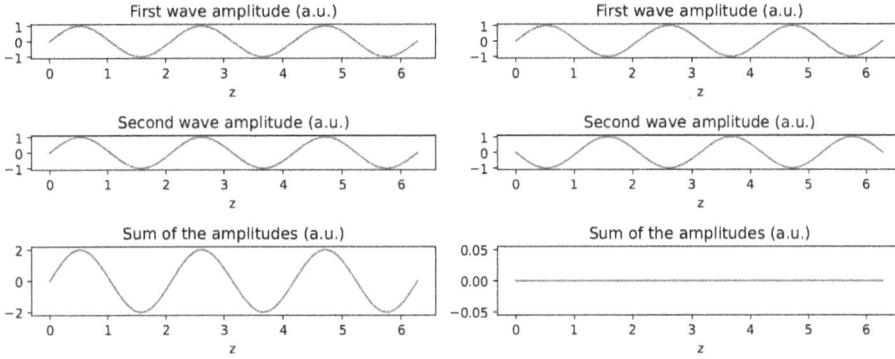

Figure 2.12. Comparison between the constructive interference and the destructive interference cases. In the left panel, the two waves' amplitudes are in phase, which leads to constructive interference when they combine. On the right, they are in phase opposition, which leads to destructive interference when they combine.

Constructive interference (see left panel of figure 2.12) happens when the cosine is $+1$, which is the case whenever its argument $\Delta\Phi$ is an even multiple of π, which corresponds to a path difference equal to:

$$\Delta\Phi = 2p\pi \text{ where } p \in \mathbb{Z}$$

which corresponds to a path difference equal to:

$$\Delta\delta = p\lambda_0 \text{ where } p \in \mathbb{Z}$$

We define the contrast of the interferences as:

$$\mathcal{C} = \frac{I_{\max} - I_{\min}}{I_{\max} + I_{\min}}$$

If the initial light beam is split into two beams of equal amplitude ($E_0 = E_{01} = E_{02}$) then the minimum intensity is 0 and the max intensity is $2KE_0^2$, namely the double of the intensity I_0 of each of the two beams taken separately. The contrast in that case is equal to 1. Otherwise, it is equal to:

$$\mathcal{C} = \frac{2E_{01}E_{02}}{E_{01}^2 + E_{02}^2} = \frac{2\sqrt{I_1}\sqrt{I_2}}{I_1 + I_2}$$

2.4.6 Spatial complex amplitude

Instead of using the electric field expression, one can use the field complex amplitude introduced in section 2.3.6. Its expression is the following:

$$\underline{E}(\vec{r}) = E_0 \exp\left(i(\vec{k}\cdot\vec{r} - \omega t + \phi)\right)$$

Under the condition that all the waves interfering have the same angular frequency, one can simplify even further by using what is called the spatial complex amplitude, for which we will write $\underline{\mathcal{E}}(\vec{r})$ here. Its expression is given by:

$$\mathcal{E}(\vec{r}) = E_0 \exp(i(\vec{k} \cdot \vec{r} + \phi)) \text{ such that } \vec{E}(\vec{r}, t) = \mathcal{E}(\vec{r})\exp(-i\omega t)$$

Thus, let us rewrite expression (2.3):

$$I = \frac{K}{2}\vec{E} \cdot \vec{E}* = \frac{K}{2}\mathcal{E}(\vec{r})\exp(-i\omega t)\mathcal{E}*(\vec{r})\exp(i\omega t)$$

which leads to:

$$I = \frac{K}{2}\mathcal{E}(\vec{r})\mathcal{E}*(\vec{r}) \tag{2.8}$$

allowing us to drop any time dependency in the calculations.

2.4.7 Young's double-slit experiment

The simplest way to obtain coherent sources is to use a single source and spatially separate the light by using an opaque plane with two holes in it. We focus here for example on the case where these holes are slits that are infinitesimally thin in one direction and infinitely long in the orthogonal direction. This configuration is called Young's double-slit experiment and is shown in figure 2.13.

The path difference between the two paths SP'M and SPM is given by:

$$\Delta\delta = \sqrt{D^2 + \left(X + \frac{b}{2}\right)^2} - \sqrt{D^2 + \left(X - \frac{b}{2}\right)^2}$$

where X is the position of point M along the x direction, b is the spacing between the slits, and D is the distance between the plane of the slits Oxy and the screen. We have:

$$\Delta\delta = D\left(\sqrt{1 + \left(\frac{X + \frac{b}{2}}{D}\right)^2} - \sqrt{1 + \left(\frac{X - \frac{b}{2}}{D}\right)^2}\right)$$

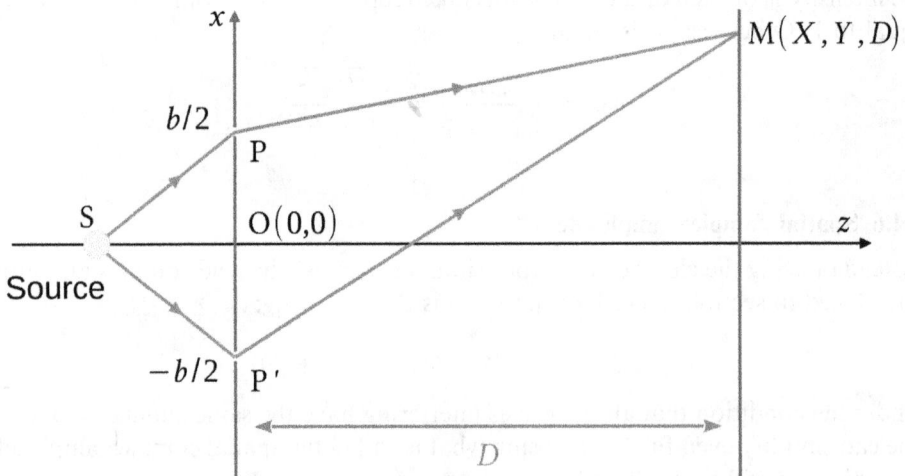

Figure 2.13. Experimental setup for Young's slits interference.

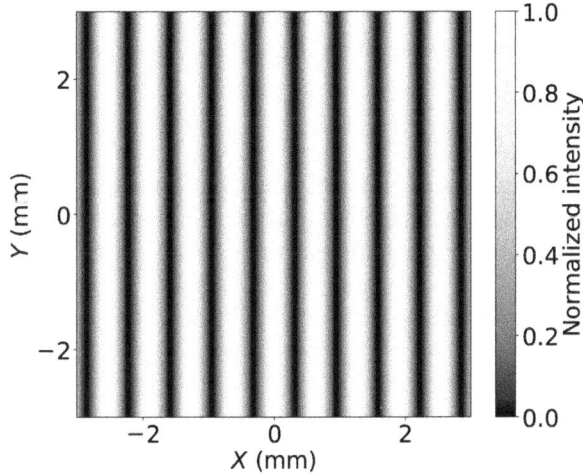

Figure 2.14. Interference pattern obtained at a distance $D = 0.5$ m from a pair of Young's slits separated by a distance $b = 0.5$ mm. The wavelength was chosen as $\lambda = 633$ nm.

A Taylor expansion for $X \ll D$ and $b \ll D$ gives:

$$\Delta\delta = D\left(1 + \frac{1}{2}\left(\frac{X + \frac{b}{2}}{D}\right)^2\right) - D\left(1 + \frac{1}{2}\left(\frac{X - \frac{b}{2}}{D}\right)^2\right)$$

Which finally gives:

$$\Delta\delta = \frac{bX}{D}$$

The intensity is given by expression (2.7):

$$I_{\text{tot}} = I_0\left(1 + \cos\left(\frac{2\pi bX}{\lambda D}\right)\right)$$

where we assume that both beams have the same intensity. The intensity is maximum for $X = p\frac{\lambda D}{b}$ with $p \in \mathbb{Z}$, which means that the interference pattern corresponds to regularly spaced linear fringes perpendicular to the x direction, as shown in figure 2.14.

2.5 Diffraction of light

We limited ourselves to two-wave interference so far. Of course, many more waves can interfere and give rise to what we call diffraction patterns [5]. The diffraction phenomenon can be a difficulty to overcome in certain cases, like when unwanted interference occurs between light waves entering an imaging apparatus (telescopes, cameras, etc.), but it can also benefit when mastered, as in spectrometers for example (see problem 2.2). In our daily life, we can observe diffraction too: for example it is diffraction that stretches the image of objects when we squint or when we look at light bulbs through curtains. Note that diffraction can occur with any type of waves, not only optical ones.

2.5.1 Huygens–Fresnel principle

It is possible to compute the light intensity distribution in a plane behind a hole with an arbitrary shape passed through by an incident plane electromagnetic wave (see figure 2.15), by combining contributions from two scientists: Huygens and Fresnel. In this section, we will derive Huygens's and Fresnel's integral, that is valid under certain conditions, and in later sections we will apply it to different masks.

2.5.1.1 Huygens's contribution

In 1690, Christian Huygens states the following principle that is now named after him: any point in space that is reached by a wave acts as a secondary source of a spherical wavelet, that has the same wavelength and phase as the incident wave. This principle explains why light can appear behind an object at places where we do not expect them, and more generally the propagation of light, as shown in figures 2.16(a)–(c).

In the case of a mask, Huygens's principle states that each infinitesimal surface element dS of the mask hole, at a position P, emits a spherical wavelet, whose amplitude is proportional to that surface dS. That wavelet is not isotropic, meaning that its amplitude also depends on the direction towards which it is emitted. Mathematically, the elementary complex amplitude of the electric field contribution that comes from dS with incident amplitude E_0 and meets a given point M at t is written as:

$$d\underline{\mathcal{E}}_P(M) = \frac{dS}{PM}K(M)E_0 \exp(i\Phi(P, M)) \text{ with } \Phi(P, M) = \vec{k} \cdot \overrightarrow{PM}$$

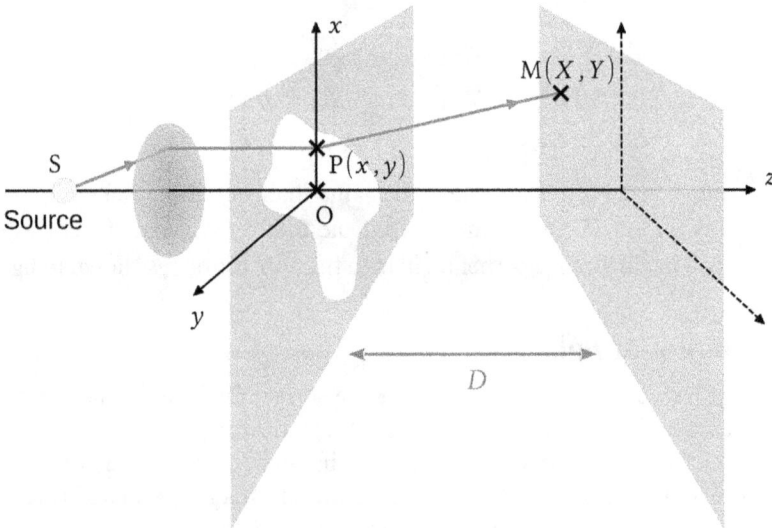

Figure 2.15. Experimental setup to observe diffraction, including the notations used later on.

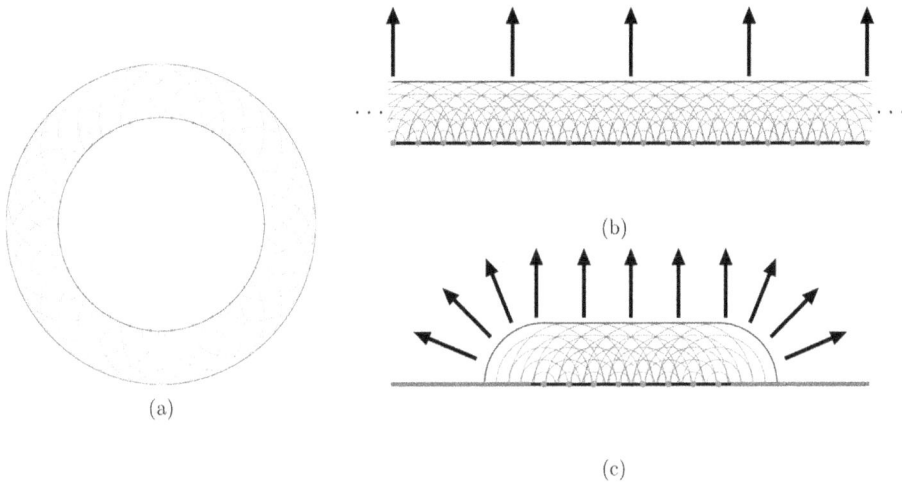

Figure 2.16. Propagation of light illustrated from the point of view of Huygens's principle, in the case of (a) a spherical wave (view in 2D), (b) a plane wave and (c) a plane wave crossing an object with a hole. The lateral wavefronts in panel (c) show the existence of beams pointing towards places where geometrical optics does not foresee any light to go.

- The term $1/\text{PM}$ shows that the light intensity evolves as an inverse square with the distance from the secondary source, as for any spherical wave.
- The component dS shows that the wave amplitude is proportional to the surface of the secondary source.
- The factor $K(\text{M})$ is called the obliquity factor and traduces that the secondary sources are not isotropic. Indeed, the intensity of the light emitted by a secondary source is proportional to $1 + \cos\alpha(\text{M})$ where $\alpha(\text{M})$ is the angle between the normal to the original wavefront at point P (which does not depend on the position of P in the case of a plane wave) and the normal to the secondary wavefront emitted towards M. Thus the intensity of the secondary wave is maximum in the direction of the incident light, whereas it is zero in the opposite direction (no light emitted towards the initial source).

2.5.1.2 Fresnel's contribution

In 1817, Fresnel associates Huygens's idea with the concept of interference that was introduced by Young. Indeed, all of the secondary sources in the mask hole are generated by the same incident wave, so they have the same angular frequency and phase jumps. Thus, as we saw in section 2.4.3, the secondary waves they emit can interfere.

Fresnel managed to accurately describe the diffraction phenomenon by deducing that the mask hole corresponds to an infinity of secondary sources emitting spherical wavelets and that the spatial complex amplitude of the electric field at point M is proportional to the sum of the complex amplitudes of all these spherical wavelets:

$$\underline{\mathcal{E}}(\text{M}) \propto \iint_{\text{P}\in \text{ hole}} d\underline{\mathcal{E}}_{\text{P}}(\text{M})$$

$$\mathcal{E}(\mathrm{M}) \propto \iint_{P\in \text{hole}} \frac{\mathrm{d}S}{\mathrm{PM}} K(\mathrm{P},\mathrm{M}) E_0(\mathrm{P}) \exp(i\vec{k} \cdot \vec{\mathrm{PM}})$$

This integral is named Huygens–Fresnel integral. It has no general analytic solution. It is thus necessary to use simplifications [6].

2.5.2 The Fresnel diffraction: rather far field approximation

The first approximation is the far field approximation, which consists in having $\|\vec{\mathrm{OP}}\| \ll \|\vec{\mathrm{OM}}\|$. It means that we observe the light intensity at a distance that is much higher than the typical dimension of the diffracting hole. We can write $\vec{\mathrm{PM}}$ as:

$$\vec{\mathrm{PM}} = \vec{\mathrm{PO}} + \vec{\mathrm{OM}} = \vec{\mathrm{OM}} - \vec{\mathrm{OP}}$$

so:

$$\|\vec{\mathrm{PM}}\| = \sqrt{(\vec{\mathrm{OM}} - \vec{\mathrm{OP}})^2} = \sqrt{\mathrm{OM}^2 - 2\,\vec{\mathrm{OM}} \cdot \vec{\mathrm{OP}} + \mathrm{OP}^2}$$

$$= \mathrm{OM}\sqrt{1 - \frac{2\,\vec{\mathrm{OM}} \cdot \vec{\mathrm{OP}} - \mathrm{OP}^2}{\mathrm{OM}^2}}$$

A Taylor expansion (see section 1.3) with $\mathrm{OP} \ll \mathrm{OM}$ gives:

$$\|\vec{\mathrm{PM}}\| = \mathrm{OM}\left(1 - \frac{2\,\vec{\mathrm{OM}} \cdot \vec{\mathrm{OP}} - \mathrm{OP}^2}{2\mathrm{OM}^2}\right) = \mathrm{OM} - \frac{2\,\vec{\mathrm{OM}} \cdot \vec{\mathrm{OP}} - \mathrm{OP}^2}{2\mathrm{OM}}$$

Let us focus on the last part of that expression:

$$\frac{2\,\vec{\mathrm{OM}} \cdot \vec{\mathrm{OP}} - \mathrm{OP}^2}{2\mathrm{OM}} \tag{2.9}$$

The first term is independent of $\|\vec{\mathrm{OM}}\|$. Indeed, it is equal to:

$$\frac{2\,\vec{\mathrm{OM}} \cdot \vec{\mathrm{OP}}}{2\mathrm{OM}} = \|\vec{\mathrm{OP}}\|\cos\widehat{\mathrm{MOP}}$$

where $\widehat{\mathrm{MOP}}$ is the angle between the vectors $\vec{\mathrm{OM}}$ and $\vec{\mathrm{OP}}$

This is much smaller than $\|\vec{\mathrm{OM}}\|$ since $\|\vec{\mathrm{OP}}\| \ll \|\vec{\mathrm{OM}}\|$ and $\cos\widehat{\mathrm{MOP}} \in [-1, 1]$.

The second term of expression (2.9) is much smaller than $\|\vec{\mathrm{OM}}\|$ as well.

As a consequence, the factor governing the modulus of the complex electric field $1/\mathrm{PM}$ can be considered as equal to $1/\mathrm{OM}$, which is independent of the position of P and can thus be taken out of the integral.

However, one cannot neglect the term (2.9) inside the exponential, since it is multiplied by $k = 2\pi/\lambda$. Then, if it varies by a quantity as small as $\lambda/2$ when P is swept over the mask hole surface, it entails a phase difference of π between two points of the mask hole, which would make the complex exponential change sign and then have a significant impact on the interference pattern.

Within Fresnel's approximation, Huygens–Fresnel's integral can be written as:

$$\mathcal{E}(\mathrm{M}) = \frac{E_0 \exp(ik\ \mathrm{OM})}{\mathrm{OM}} \iint_{\mathrm{P} \in \text{ hole}} \mathrm{d}S K(\mathrm{P},\ \mathrm{M}) \exp\left(-ik\frac{2\,\overrightarrow{\mathrm{OM}} \cdot \overrightarrow{\mathrm{OP}} - \mathrm{OP}^2}{2\,\mathrm{OM}}\right).$$

2.5.3 Fraunhofer diffraction

2.5.3.1 Even further field approximation
We saw in the previous section that one cannot simply neglect the term (2.9) in the exponential. However, by making a more drastic assumption about how big OM is, one can simplify Huygens–Fresnel's expression a bit more. No matter how large OM is, the first part of that term ($\overrightarrow{\mathrm{OM}} \cdot \overrightarrow{\mathrm{OP}}/\mathrm{OM}$) cannot be neglected since it does not depend on it. On the other hand, the second part, $\mathrm{OP}^2/(2\,\mathrm{OM})$, depends on $\|\overrightarrow{\mathrm{OM}}\|$. The Fraunhofer approximation consists in considering that the screen on which the diffraction pattern is observed is sufficiently far away from the diffracting element that:

$$\frac{\mathrm{OP}^2}{2\mathrm{OM}} \ll \frac{\lambda}{2\pi}$$

Within the Fraunhofer approximation, Huygens–Fresnel's integral can be written as:

$$\mathcal{E}(\mathrm{M}) = \frac{E_0 \exp(ik\ \mathrm{OM})}{\mathrm{OM}} \iint_{\mathrm{P} \in \text{ hole}} \mathrm{d}S K(\mathrm{P}, \mathrm{M}) \exp\left(-ik\frac{\overrightarrow{\mathrm{OM}} \cdot \overrightarrow{\mathrm{OP}}}{\mathrm{OM}}\right) \quad (2.10)$$

2.5.3.2 Paraxial approximation
This approximation consists in considering that the optical beams are almost parallel to the optical axis. Mathematically, the paraxial approximation translates as $X, Y \ll D$ where $(X, Y, Z = D)$ are the cartesian coordinates of point M, and $x, y \ll D$ where $(x, y, z = 0)$ are the cartesian coordinates of the secondary wave (point of the diffracting element). Within that approximation, the obliquity factor can be considered constant equal to $1 + \cos(0) = 2$. That value is often disregarded and simply considered equal to 1.

Since at most $\mathrm{OP} = e$ with e the typical length of the diffracting element in any direction, and $\mathrm{OM} \simeq D$ in the paraxial approximation $((X, Y, Z = D)$ being the coordinates of point M), we can rewrite the condition for Fraunhofer approximation as:

$$D \gg \frac{\pi e^2}{\lambda}$$

That means that to observe Fraunhofer diffraction with a red light ($\lambda \sim 630$ nm) with a hole of typical size $e = 1$ mm, one has to place the screen at at least $D = \pi \times (1 \times 10^{-3})^2/(630 \times 10^{-9}) \simeq 5$ m from the hole.

2.5.4 Diffraction in slits and apertures in the Fraunhofer approximation

Let us apply Huygens's and Fresnel's integral to different cases.

2.5.4.1 Single slit

Let us consider first the case of a single slit of width a in the x direction, and length L in the y direction, with $L>>a$. In the limit where L tends to infinity, we can assume that the light intensity repartition is invariant along the y direction. That means that the phase of each wave at position M does not depend on the variable Y. The surface element dS can thus be expressed as Ldx.

$$\underline{\mathcal{E}}(M) = \int_{x=-a/2}^{x=+a/2} d\underline{\mathcal{E}}(M)$$

In this case, equation (2.10) becomes:

$$\underline{\mathcal{E}}(M) = \frac{E_0 \exp(ik \text{ OM })}{\text{OM}} \int_{x=-a/2}^{x=+a/2} \exp(-ikx \sin \theta)\, Ldx$$

where we used that $\overrightarrow{\text{OM}} \cdot \overrightarrow{\text{OP}}/\text{OM} = \text{OM}$. OP $\cos(\pi/2 - \theta)/\text{OM} = x \sin \theta$ (see figure 2.17).

$$\int_{x=-a/2}^{x=+a/2} \exp(-ikx \sin \theta)\, dx = \frac{1}{-ik \sin \theta}[\exp(-ikx \sin \theta)]_{x=-a/2}^{x=+a/2}$$

$$= \frac{1}{-ik \sin \theta}\left(\exp\left(-ik\frac{a}{2} \sin \theta\right) - \exp\left(+ik\frac{a}{2} \sin \theta\right)\right)$$

$$= \frac{1}{-ik \sin \theta}\left(-2i \sin\left(k\frac{a}{2} \sin \theta\right)\right) = a\frac{\sin\left(k\frac{a}{2} \sin \theta\right)}{k\frac{a}{2} \sin \theta} \quad (2.11)$$

$$= a \text{ sinc}\left(k\frac{a}{2} \sin \theta\right)$$

The corresponding intensity is obtained using equation (2.8):

$$I(M) \propto \underline{\mathcal{E}}(M) \cdot \underline{\mathcal{E}}(M)^* = \frac{E_0^2 \cos^2 \theta}{D^2}L^2a^2 \text{ sinc }^2\left(k\frac{a}{2} \sin \theta\right)$$

where we used that $\text{OM} = D/\cos \theta$ where D is the distance between the slit and the screen.

Thus, we observe the pattern shown in figure 2.18(a). The intensity as a function of the variable θ (defined in the Oxz plane, see figure 2.17) is shown in figure 2.18(b). The intensity cancels for:

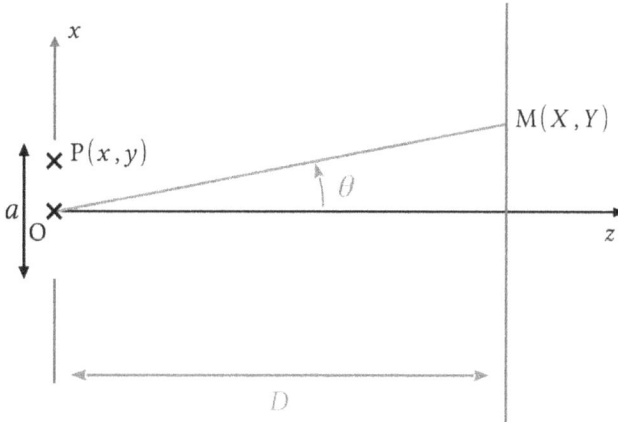

Figure 2.17. Drawing of diffraction by an infinite single slit, including the notations used in the main text.

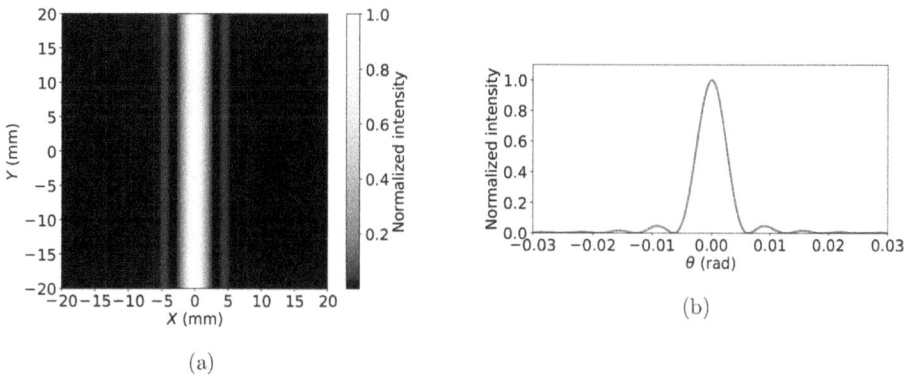

(a)

(b)

Figure 2.18. (a) Intensity pattern on the screen resulting from diffraction of light by an infinitely long single slit. (b) Intensity profile as a function of θ. See setup drawing in figure 2.17 for the corresponding notations. The parameters were chosen as $\lambda = 633$ nm, $a = 100$ μm and $D = 0.5$ m.

$$k\frac{a}{2}\sin\theta = p\pi \text{ with } p \in \mathbb{Z}^* \text{ or } \boxed{\sin\theta = p\frac{\lambda}{a} \text{ with } p \in \mathbb{Z}^*} \tag{2.12}$$

Figure 2.19 gives a schematic interpretation of the phenomenon [7]. At zero angle, the waves coming from all of the secondary sources belonging to the slit interfere constructively. Now if we look at the wavelets going towards a higher point, with a small angle $\theta > 0$, we see that some of the wavelets (coloured in pink) see their contribution cancelled by those from other wavelets (coloured in green). As θ increases, this cancellation impacts a bigger and bigger proportion of the wavelets, until the intensity reaches zero when exactly half of the wavelets amplitudes get cancelled by those of the other half. This is the first intensity minimum, occurring for $\sin\theta = \lambda/a$. As θ keeps increasing, the proportion of wavelets whose contribution is cancelled decreases below half, leading to a new increase of intensity. The next cancellations happen when $\sin(\theta) = p\lambda/a$ with $p = 2, 3, \ldots$ as the calculation showed before, see expression (2.12).

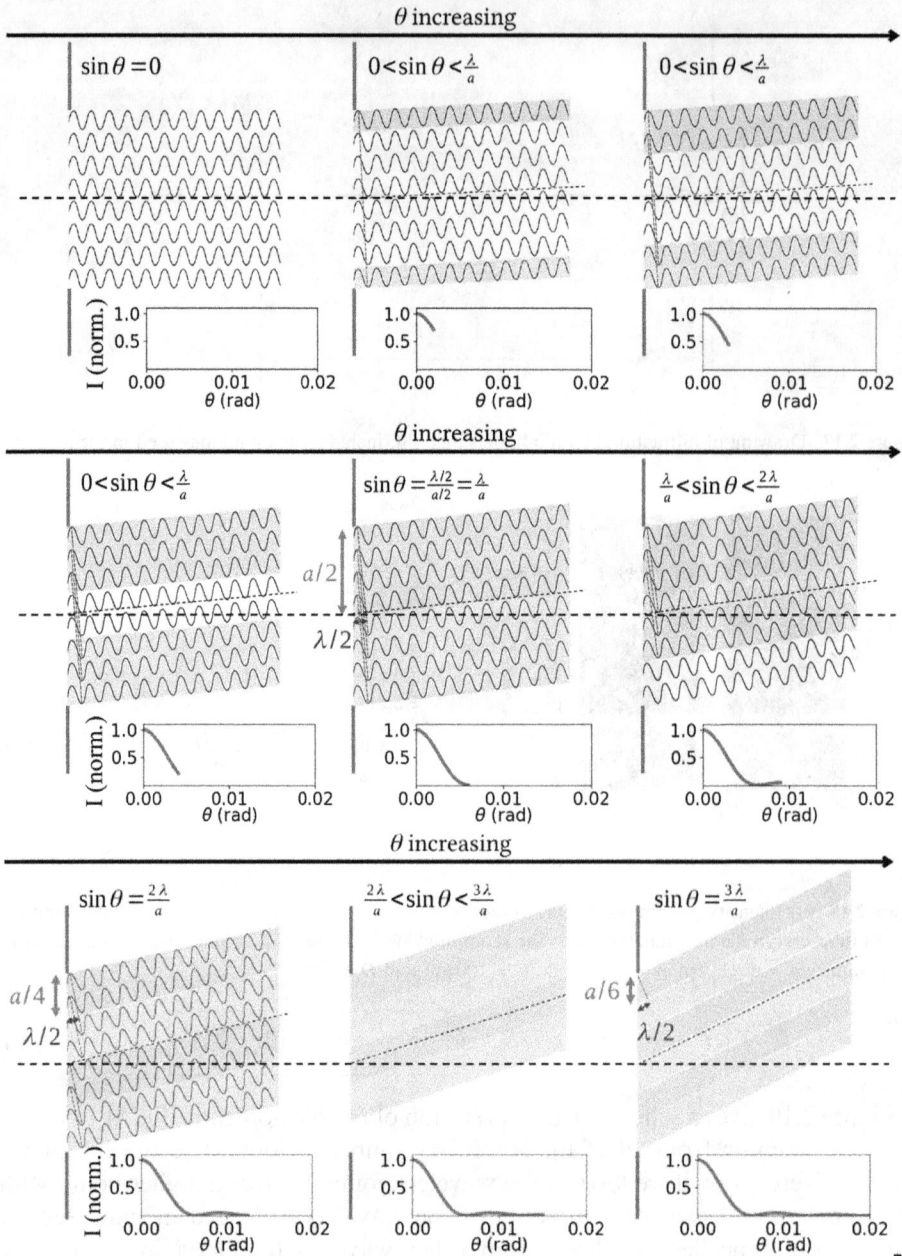

Figure 2.19. Interpretation of the diffraction by a single slit.

2.5.4.2 Double slit

Let us now consider the case of two slits of width a in the x direction and length L in the y direction with $L>>a$, and spaced by a distance b. Such a setup is represented in

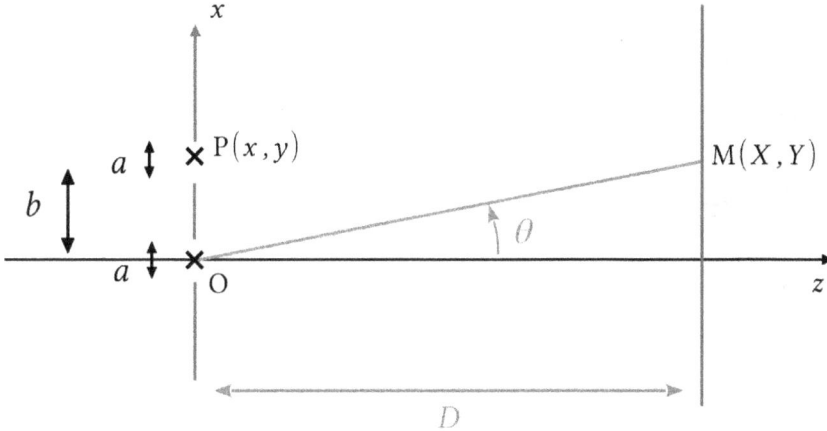

Figure 2.20. Drawing of diffraction by two infinite slits, including the notations used in the main text.

figure 2.20. Here again, we can assume that the light intensity repartition is invariant along the y direction.

$$\mathcal{E}(M) = \int_{x=-a/2}^{x=+a/2} d\mathcal{E}(M) + \int_{x=b-a/2}^{x=b+a/2} d\mathcal{E}(M)$$

$$\mathcal{E}(M) = \frac{E_0 \exp(ik\ OM)L}{OM}\left(\int_{x=-a/2}^{x=+a/2} \exp(-ikx \sin \theta)\, dx + \int_{x=b-a/2}^{x=b+a/2} \exp(-ikx \sin \theta)\, dx\right)$$

since the expression of the elementary spatial complex amplitude is the same as in the case of the single slit.

By performing calculations similar to those presented in equation (2.11), we obtain the following expressions:

$$\int_{x=-a/2}^{x=+a/2} \exp(-ikx \sin \theta)\, dx = a\ \mathrm{sinc}\left(k\frac{a}{2} \sin \theta\right)$$

$$\int_{x=b-a/2}^{x=b+a/2} \exp(-ikx \sin \theta)\, dx = a \exp(-ikb \sin \theta)\ \mathrm{sinc}\left(k\frac{a}{2} \sin \theta\right)$$

(2.13)

The corresponding intensity is obtained using equation (2.8):

$$I(M) \propto \mathcal{E}(M) \cdot \mathcal{E}(M)^* = \frac{2E_0^2 \cos^2 \theta}{D^2}(1 + \cos(kb \sin \theta))L^2 a^2\ \mathrm{sinc}^2\left(k\frac{a}{2} \sin \theta\right)$$

where we used again that $OM = D/\cos \theta$ where D is the distance between the plane of the slits and the screen. We recognize the intensity repartition given by Young's slits, but multiplied by the intensity repartition of a single slit.

Thus, we observe the pattern shown in figure 2.21(a). The intensity as a function of the variable θ (defined in the Oxz plane, see figure 2.20) is shown in figure 2.21(b).

We will study the case of N slits in a problem at the end of this chapter (problem 2.2).

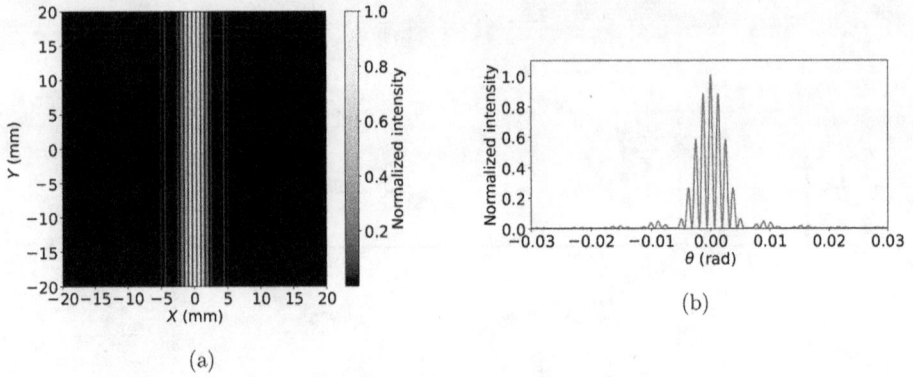

Figure 2.21. (a) Intensity pattern on the screen resulting from diffraction of light by two parallel infinitely long slits. (b) Intensity profile as a function of θ. See setup drawing in figure 2.20 for the corresponding notations. The parameters were chosen as $\lambda = 633$ nm, $a = 100\,\mu$m, $b = 500\,\mu$m and $D = 0.5$ m.

2.5.4.3 Rectangular aperture

In the case of a rectangular aperture, the complex amplitude is the following:

$$\mathcal{E}(M) = \frac{E_0 \exp(ik\,OM)}{OM} \int_{x=-a/2}^{x=+a/2} \int_{y=-b/2}^{y=+b/2} \exp(-ik(x\sin\theta_x + y\sin\theta_y))dxdy$$

where a and b represent the size of the slit along the x and y axes, respectively. We used that:

$$\frac{\overrightarrow{OM} \cdot \overrightarrow{OP}}{OM} = \frac{\overrightarrow{OM} \cdot (x\overrightarrow{u_x} + y\overrightarrow{u_y})}{OM} = \frac{OM\,x\cos\left(\frac{\pi}{2} - \theta_x\right) + OM\,y\cos\left(\frac{\pi}{2} - \theta_y\right)}{OM}$$
$$= x\sin\theta_x + y\sin\theta_y \tag{2.14}$$

where $\overrightarrow{u_x}$ (resp. $\overrightarrow{u_y}$) is the unitary vector along the x (resp. y) axis.

The integral can be split in two, one over x and one over y. Then, using again the result for a single slit, namely equation (2.11), we have:

$$\int_{x=-a/2}^{x=+a/2} \exp(-ikx\sin\theta_x)\,dx = a\,\text{sinc}\left(k\frac{a}{2}\sin\theta_x\right)$$

and similarly:

$$\int_{y=-b/2}^{y=+b/2} \exp(-iky\sin\theta_y)\,dy = b\,\text{sinc}\left(k\frac{b}{2}\sin\theta_y\right)$$

Thus:

$$\mathcal{E}(M) = \frac{E_0 \exp(ik\,OM)}{OM}ab\,\text{sinc}\left(k\frac{a}{2}\sin\theta_x\right)\text{sinc}\left(k\frac{b}{2}\sin\theta_y\right) \tag{2.15}$$

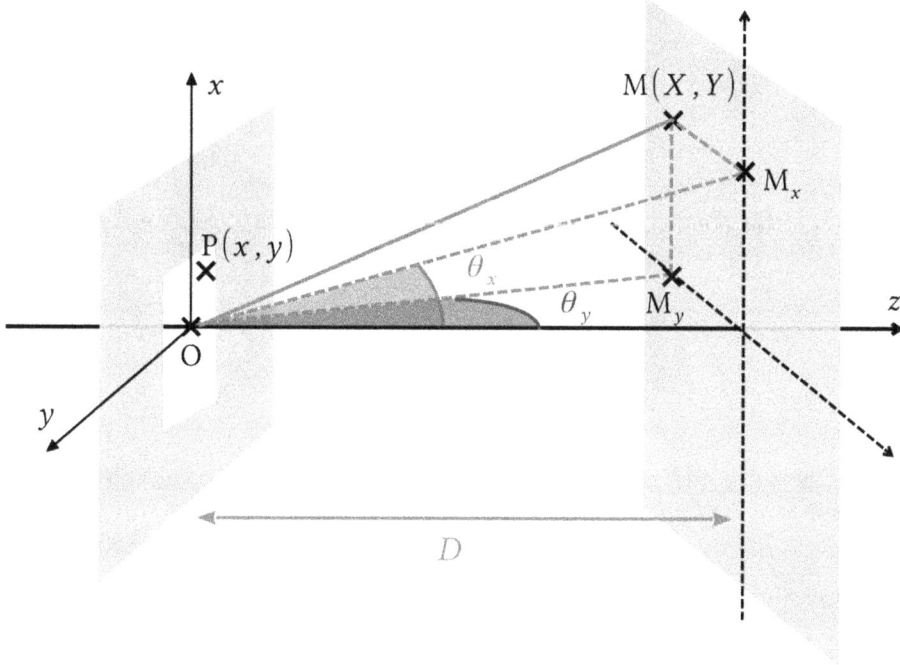

Figure 2.22. Drawing of diffraction by a rectangular aperture, including the notations used in the main text.

The corresponding intensity is obtained using equation (2.8):

$$I(\mathrm{M}) \propto \underline{\mathcal{E}}(\mathrm{M}) \cdot \underline{\mathcal{E}}(\mathrm{M})^* = \left(\frac{E_0 \cos \theta_x \cos \theta_y ab}{D} \right)^2 \mathrm{sinc}^2 \left(k \frac{a}{2} \sin \theta_x \right) \mathrm{sinc}^2 \left(k \frac{b}{2} \sin \theta_y \right)$$

where we used the following trigonometric relationships, that appear from figure 2.22:

$$\cos \theta_y = \frac{\mathrm{OM}_x}{\mathrm{OM}} \text{ where } \mathrm{M}_x \text{ is the projection of M on the } x\text{-axis}$$

$$\cos \theta_x = \frac{D}{\mathrm{OM}_x}$$

$$\text{so } \mathrm{OM} = \frac{D}{\cos \theta_x \cos \theta_y}$$

Finally, the intensity in the screen plane is shown in figure 2.23.

2.5.4.4 Circular hole

Following the cylindrical symmetry of this situation, it is more convenient to use polar coordinates here. We use the notations displayed in figure 2.24. We have the following trigonometric relationships:

$$\sin \theta = \frac{\mathrm{O'M}}{\mathrm{OM}} \text{ and } \cos \phi' = \frac{X}{\mathrm{O'M}} \text{ and } \sin \phi' = \frac{Y}{\mathrm{O'M}}$$

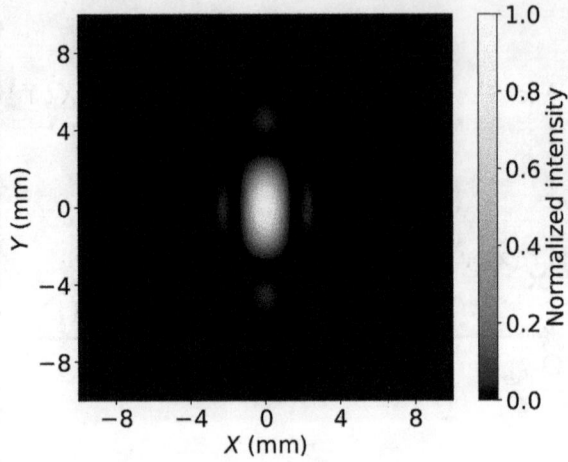

Figure 2.23. Diffraction figure obtained from a rectangular aperture. The parameters used for that figure are $a = 200\,\mu\text{m}$, $b = 100\,\mu\text{m}$, $E_0 = 1$ arbitrary unit, $\lambda = 633\,\text{nm}$ and $D = 50\,\text{cm}$.

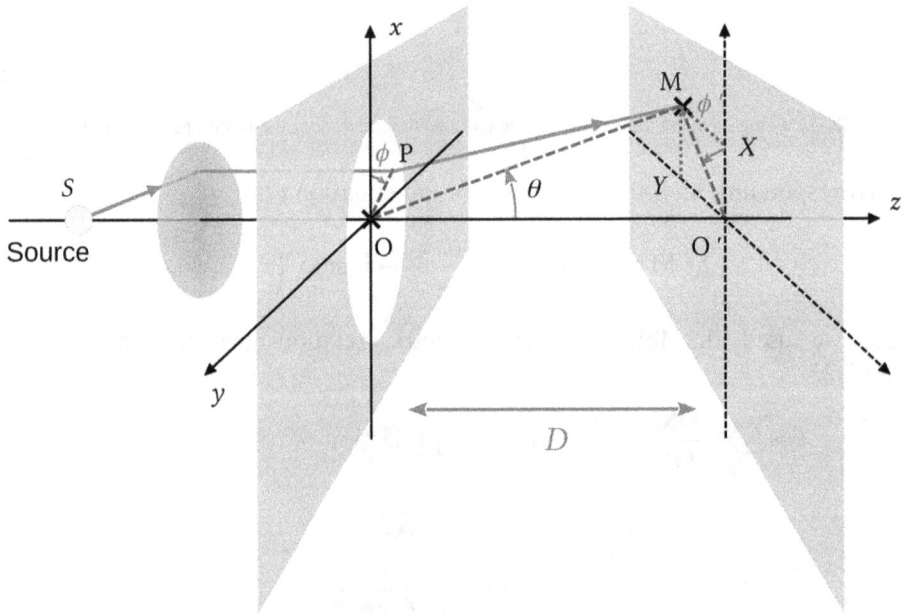

Figure 2.24. Drawing of the diffraction of light through a circular hole.

We thus have the following coordinates for vectors $\overrightarrow{\text{OM}}$ and $\overrightarrow{\text{OP}}$:

$$\overrightarrow{\text{OM}} = \begin{pmatrix} \text{O'M}\cos\phi' \\ \text{O'M}\sin\phi' \\ D \end{pmatrix} = \begin{pmatrix} \text{OM}\sin\theta\cos\phi' \\ \text{OM}\sin\theta\sin\phi' \\ D \end{pmatrix} \text{ and } \overrightarrow{\text{OP}} = \begin{pmatrix} \text{OP}\cos\phi \\ \text{OP}\sin\phi \\ 0 \end{pmatrix}$$

Finally:

$$\frac{\overrightarrow{OM} \cdot \overrightarrow{OP}}{OM} = OP \sin\theta \cos\phi' \cos\phi + OP \sin\theta \sin\phi' \sin\phi$$

where we can use the trigonometric formula (1.3) and obtain:

$$\frac{\overrightarrow{OM} \cdot \overrightarrow{OP}}{OM} = OP \sin\theta \cos(\phi - \phi')$$

We can now insert that expression into the Huygens–Fresnel's integral:

$$\underline{\mathcal{E}}(M) = \frac{E_0 \exp(ik\ OM)}{OM} \int_{OP=0}^{OP=R} \int_{\phi=0}^{\phi=2\pi} \exp(-ik\ OP \sin\theta \cos(\phi - \phi'))\ OPdOPd\phi$$

where R is the hole radius. ϕ' is fixed for a given position M so it is a constant in the integral, and cos being 2π periodic, one can show (see problem 2.1) that the result is independant of ϕ'. Thus, for simplification, let us set the value of ϕ' to zero:

$$\underline{\mathcal{E}}(M) = \frac{E_0 \exp(ik\ OM)}{OM} \int_{OP=0}^{OP=R} \underbrace{\int_{\phi=0}^{\phi=2\pi} \exp(-ik\ OP \sin\theta \cos\phi)d\phi}_{2\pi J_0(-k\ OP \sin\theta)}\ OPdOP$$

where one recognizes the Bessel function $J_m(x)$ defined in equation (1.17) for $m = 0$. We then implement the following change of variable: $x = -kOP \sin\theta$:

$$\underline{\mathcal{E}}(M) = \frac{2\pi E_0 \exp(ik\ OM)}{OM\ k^2 \sin^2\theta} \underbrace{\int_{x=0}^{x=-kR\sin\theta} x J_0(x)\ dx}_{-kR\sin\theta J_1(-kR\sin\theta)}$$

where we used a property of the Bessel functions introduced in equation (1.16). To compute the light intensity repartition from this expression of the complex electric field, we use the usual formula (2.18):

$$I(M) \propto \underline{\mathcal{E}}(\overrightarrow{r}, t)\underline{\mathcal{E}}^*(\overrightarrow{r}, t) = \left(\frac{2\pi E_0 R}{Dk \tan\theta} J_1(-kR \sin\theta)\right)^2$$

where we used that $\cos\theta = D/OM$ where D is the distance between the hole and the observation screen. J_1 is an odd function since $m = 1$ is odd, then $J_1(-kR \sin\theta) = -J_1(kR \sin\theta)$, thus the final result is:

$$I(M) \propto \left(\frac{E_0 R}{Dk \tan\theta} J_1(kR \sin\theta)\right)^2$$

The obtained diffraction pattern is shown on figure 2.25(a). To better illustrate the behaviour of the intensity, it is also plotted as a function of $\overset{'}{O}M$ in figure 2.25(b).

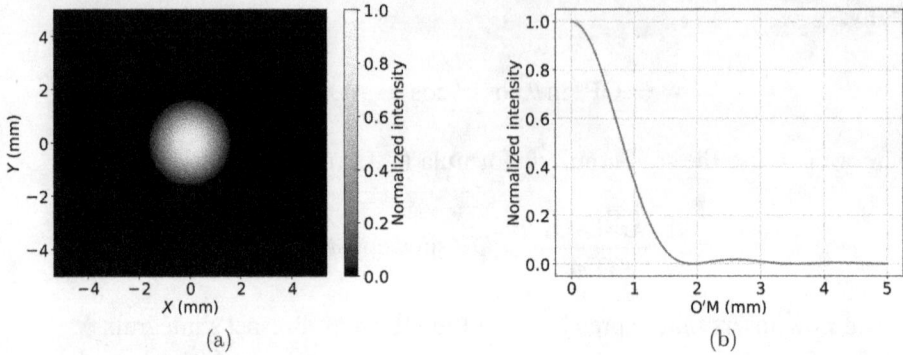

Figure 2.25. (a) Intensity pattern on the screen resulting from diffraction of light by a circular hole. (b) Intensity profile as a function of $O'M$. See setup drawing in figure 2.24 for the corresponding notations. The parameters were chosen as $\lambda = 600\,\text{nm}$, $R = 100\,\mu\text{m}$ and $D = 0.5\,\text{m}$.

2.6 Problems

Problem 2.1: Invariance of the diffraction figure for a circular hole by rotation around the optical axis

To show the invariance of the diffraction figure obtained for a circular hole by rotation around the optical axis, show that:

$$\int_{\phi=0}^{\phi=2\pi} \exp(-ik\,\text{OP}\,\sin\theta\,\cos(\phi - \phi'))\,\mathrm{d}\phi = \int_{\phi=0}^{\phi=2\pi} \exp(-ik\,\text{OP}\,\sin\theta\,\cos(\phi))\,\mathrm{d}\phi$$

Solution:

Let us change the variable as follows: $\phi'' = \phi - \phi'$. Then, as ϕ evolves from 0 to 2π, ϕ'' evolves from $-\phi'$ to $2\pi - \phi'$. Thus the quantity we intend to calculate is written:

$$Q = \int_{\phi=0}^{\phi=2\pi} \exp(-ik\,\text{OP}\,\sin\theta\,\cos(\phi - \phi'))\,\mathrm{d}\phi = \int_{\phi''=-\phi'}^{\phi''=2\pi-\phi'} \exp(-ik\,\text{OP}\,\sin\theta\,\cos(\phi''))\,\mathrm{d}\phi''$$

We can then use the additivity property of integrals (see equation (1.8)):

$$\int_{\phi''=-\phi'}^{\phi''=2\pi-\phi'} \exp\left(-ik\,\text{OP}\,\sin\theta\,\cos(\phi'')\right)\,\mathrm{d}\phi'' = \int_{\phi''=-\phi'}^{\phi''=0} \exp\left(-ik\,\text{OP}\,\sin\theta\,\cos(\phi'')\right)\,\mathrm{d}\phi''$$
$$+ \int_{0}^{\phi''=2\pi-\phi'} \exp\left(-ik\,\text{OP}\,\sin\theta\,\cos(\phi'')\right)\,\mathrm{d}\phi''$$

We are going to manipulate the left term of that sum. We know that the cosine function is 2π-periodic, which translates as $\cos(\alpha + 2\pi) = \cos\alpha\ \forall\ \alpha \in \mathbb{R}$. Then:

$$\int_{\phi''=-\phi'}^{\phi''=0} \exp(-ik\,\text{OP}\,\sin\theta\,\cos(\phi''))\,\mathrm{d}\phi'' = \int_{\phi''=-\phi'}^{\phi''=0} \exp(-ik\,\text{OP}\,\sin\theta\,\cos(\phi'' + 2\pi))\,\mathrm{d}\phi''$$

Let us implement another change of variable for that latter term: $\phi''' = \phi'' + 2\pi$. This leads to:

$$Q = \int_{\phi'''=2\pi-\phi'}^{\phi'''=2\pi} \exp(-ik \text{ OP} \sin\theta \cos(\phi''')) \, d\phi''' + \int_0^{\phi''=2\pi-\phi'} \exp(-ik \text{ OP} \sin\theta \cos(\phi'')) \, d\phi''$$

Using the additivity property of integrals again, we obtain what we were trying to show, which is:

$$Q - \int_0^{\phi=2\pi} \exp(-ik \text{ OP} \sin\theta \cos(\phi)) \, d\phi$$

We can then conclude that the diffraction figure obtained from a circular hole is invariant by varying ϕ' as defined in figure 2.24, as one could expect from the problem symmetries.

Problem 2.2: Diffraction by N slits and application to spectroscopy

Let us now consider the case of N slits (a 'grating'), of width a in the x direction and length L in the y direction with $L \gg a$, and regularly spaced by a distance b. Such a setup is represented in figure 2.26. A plane wave arrives on the mask.

1. Show that the intensity distribution $I(M)$ (on the screen that is placed at a distance D after the mask) displays peaks whose position depend on the light wavelength.

2. (a) Show that the different spectral components of the source can appear many times. A given series of peaks is called an order. Show that orders p and $p+1$ overlap when:

$$p \geqslant \frac{\lambda_{min}}{\lambda_{max} - \lambda_{min}}$$

where λ_{min} and λ_{max} are, respectively, the minimal and maximal wavelengths of the incident light spectrum.

(b) Determine the minimum order for which overlap is observed with the visible part of the light emitted by a mercury vapour lamp, for which the wavelengths are around 405 nm, 436 nm, 546 nm and 578 nm [8].

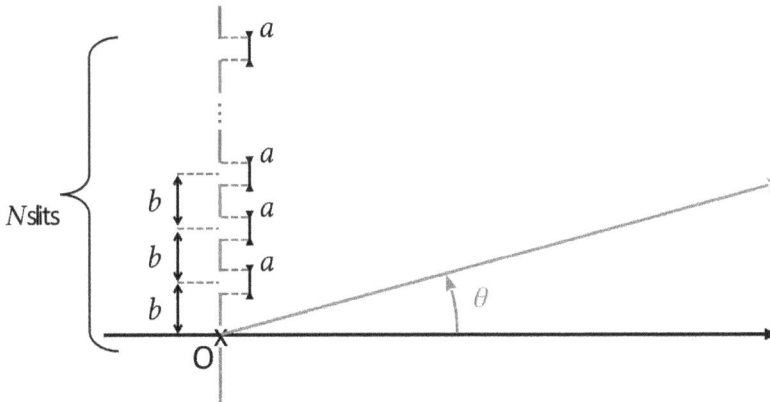

Figure 2.26. Drawing of diffraction by N infinite single slits, including the notations used in the main text.

3. Each peak of a given colour is surrounded by wings on both sides. Show that the intensity of each main peak is much bigger than that of their wings, and determine by what ratio. We will make the hypothesis that the number of slits through which the light passes is high.

4. Show, for a given wavelength, that when the number of slits N receiving light increases, the width of the peaks decreases. Show also that the bigger the observation angle θ is, the wider the peaks are.

Solution:

1.

$$\mathcal{E}(M) = \int_{x=-a/2}^{x=+a/2} d\,\mathcal{E}(M) + \int_{x=b-a/2}^{x=b+a/2} d\,\mathcal{E}(M) + \int_{x=2b-a/2}^{x=2b+a/2} d\,\mathcal{E}(M) + \cdots$$

$$= \sum_{p=0}^{N-1} \int_{x=pb-a/2}^{x=pb+a/2} d\,\mathcal{E}(M)$$

$$= \frac{E_0 \exp(ik\,OM)L}{OM} \sum_{p=0}^{N-1} \int_{x=pb-a/2}^{x=pb+a/2} \exp(-ikx \sin\theta)\,dx$$

Generalizing the results of equation (2.13), we have:

$$\int_{x=pb-a/2}^{x=pb+a/2} \exp(-ikx \sin\theta)\,dx = a \exp(-ikpb \sin\theta) \operatorname{sinc}\left(k\frac{a}{2}\sin\theta\right)$$

Thus one gets:

$$\mathcal{E}(M) = \frac{E_0 \exp(ik\,OM)}{OM} La \operatorname{sinc}\left(k\frac{a}{2}\sin\theta\right)\sum_{p=0}^{N-1}\exp(-ikpb \sin\theta)$$

We recognize the sum of a geometric progression:

$$\sum_{p=0}^{N-1}\exp(-ikpb \sin\theta) = \sum_{p=0}^{N-1}(\exp(-ikb \sin\theta))^p = \frac{1 - (\exp(-ikb \sin\theta))^N}{1 - \exp(-ikb \sin\theta)}$$

The corresponding intensity is obtained using equation (2.8):

$$I(M) \propto \mathcal{E}(M) \cdot \mathcal{E}(M)^* = \frac{E_0^2 \cos^2\theta}{D^2} L^2 a^2 \operatorname{sinc}^2\left(k\frac{a}{2}\sin\theta\right)\frac{\sin^2\left(\dfrac{kbN \sin\theta}{2}\right)}{\sin^2\left(\dfrac{kb \sin\theta}{2}\right)}$$

where we used that $OM = D/\cos\theta$ where D is the distance between the plane of the slits and the screen.

That intensity expression shows peaks where the denominator cancels:

$$\sin^2\left(\frac{kb \sin\theta}{2}\right) = 0 \iff \frac{kb \sin\theta_p}{2} = p\pi \text{ where } p \in \mathbb{Z}$$

where $\mathbb{Z}$ is the ensemble of relative integers. Finally, the condition to have a peak of intensity for a given wavelength λ is the following:

$$\sin\theta_p = p\frac{\lambda}{b} \text{ where } p \in \mathbb{Z} \tag{2.16}$$

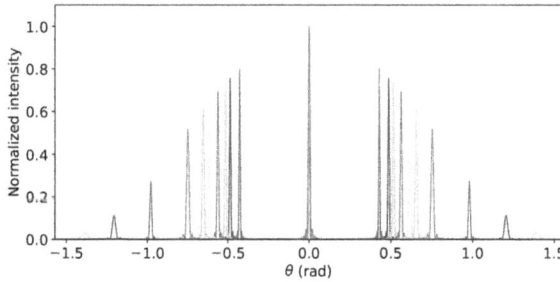

Figure 2.27. Intensity profile as a function of θ. See setup drawing in figure 2.26 for the corresponding notations. The parameters were chosen as $\lambda = 415$ nm, 467 nm, 492 nm, 532 nm, 577 nm, 607 nm, 682 nm (resp. purple, blue, cyan, green, yellow, orange and red) for the incident light wavelength, $a = 0.1$ μm for the slits width, $b = 1$ μm for the spacing between the slits, $N = 50$ for the number of slits through which the light passes, and $D = 0.5$ m.

We see that for other orders than $p = 0$, the intensity peaks corresponding to different wavelengths happen for different angles θ_p. This is illustrated in figure 2.27. Visually, intensity peaks corresponding to different wavelengths happening for different angles mean that the colour of the object depends on its structure and on the angle under which it is observed. This phenomenon occurs for various objects in our daily life: some butterflies show these structural colours, for example the Morpho butterfly [9], some shells, peacock feathers, some sparkly images or stickers (called *holographic stickers* although it is not exactly related to holography), etc.

2. There is an overlap between orders p and $p + 1$ when the smallest wavelength λ_{min} of order $p + 1$ has a peak for an angle θ that is smaller than the biggest wavelength λ_{max} of order p. Mathematically speaking, this translates as:

$$\sin(\theta_{p+1})_{\lambda_{min}} \leq \sin(\theta_p)_{\lambda_{max}} \Leftrightarrow (p + 1)\frac{\lambda_{min}}{b} \leq p\frac{\lambda_{max}}{b}$$

where we used equation (2.16). Simplifying and reorganizing that last condition, we get:

$$p(\lambda_{max} - \lambda_{min}) \geq \lambda_{min} \text{ then } p \geq \frac{\lambda_{min}}{\lambda_{max} - \lambda_{min}}$$

For the visible emission of a mercury vapour lamp, that corresponds to:

$$p \geq \frac{405}{578 - 405} \simeq 2.3$$

which means that the overlap happens from order 3 on. The intensities of the different visible spectral components of the mercury vapour lamp are plotted in figure 2.28, with the corresponding observed spectrum. One can check that the fourth purple peak (excluding the central one) arises for lower angles than the third yellow and green peaks.

3. For one given wavelength, for example the purple one (at 415 nm), the intensity as a function of θ is displayed in figure 2.29(a), together with its envelope (which is the sinc squared).

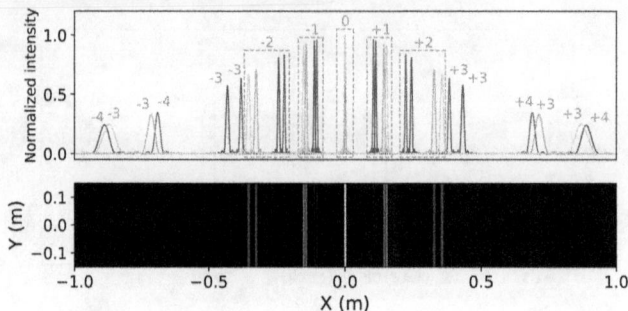

Figure 2.28. The top panel shows the intensity profile obtained from the diffraction by N slits for the spectral components of a mercury vapour lamp, as a function of the spatial coordinate X. See setup drawing in figure 2.26 for the corresponding notations. The parameters were chosen as $a = 0.01$ μm for the slits width, $b = 2$ μm for the spacing between the slits, $N = 20$ for the number of slits through which the light passes. The diffraction orders (p in the main text) are displayed in grey. The bottom panel corresponds to the actual spectrum as it would be observed after diffraction of light from a mercury vapour lamp, on a screen that is positioned at a distance $D = 0.5$ m from the grating.

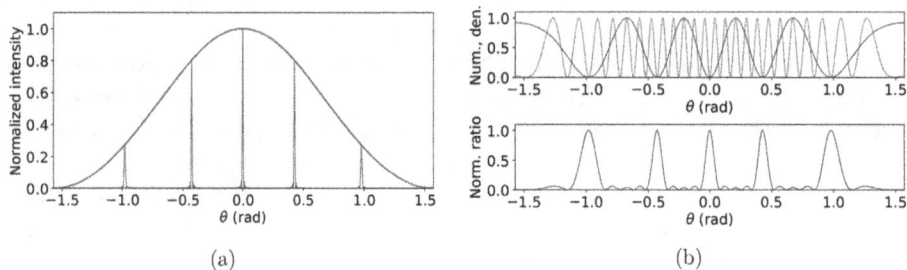

(a)   (b)

Figure 2.29. (a) Intensity profile obtained from the diffraction by N slits for one given wavelength $\lambda = 415$ nm, as a function of θ. See setup drawing in figure 2.26 for the corresponding notations. The parameters were chosen as $a = 0.1$ μm for the slits width, $b = 1$ μm for the spacing between the slits, $N = 50$ for the number of slits through which the light passes, (b) The top panel shows the normalized numerator (in red) of the intensity expression and its normalized denominator (in blue), while the bottom panel shows the ratio numerator/denominator. The parameters are the same as for figure (a), but the number of slits N has been reduced to 5 for the sake of simplicity.

The ratio between the intensity of one of the main peaks and the intensity of its immediate right neighbour (or right "wing") is:

$$R = \frac{I_{p\,\text{neighbour}}}{I_p} \text{ where } \begin{cases} I_{p\,\text{neighbour}} \propto \dfrac{\sin^2\left(\dfrac{Nkb}{2}\sin\theta_{p\,\text{neighbour}}\right)}{\sin^2\left(\dfrac{kb}{2}\sin\theta_{p\,\text{neighbour}}\right)} \\[2em] I_p \propto \dfrac{\sin^2\left(\dfrac{Nkb}{2}\sin\theta_p\right)}{\sin^2\left(\dfrac{kb}{2}\sin\theta_p\right)} \end{cases}$$

2-34

if we neglect the evolution of the envelope when θ evolves from θ_p to $\theta_{p \text{ neighbour}}$.

• Let us first compute I_p. The sine square function being π-periodic, we know that the behaviour of the numerator is the same around argument $Nkb \sin \theta_p / 2$ as around $Nkb \sin \theta_0 / 2$ which is zero, and that the behaviour of the denominator of I_p is the same around $kb \sin \theta_p / 2$ as around $kb \sin \theta_0 / 2$ which is also zero. We can thus use expression (1.7):

$$\sin^2 \left(\frac{Nkb}{2} \sin \theta_p \right) \simeq \left(\frac{Nkb}{2} \sin \theta_p \right)^2$$

$$\sin^2 \left(\frac{kb}{2} \sin \theta_p \right) \simeq \left(\frac{kb}{2} \sin \theta_p \right)^2$$

From which we get:

$$I_p \propto N^2$$

• Let us now compute $I_{p \text{ neighbour}}$. The immediate neighbour corresponds to the second next peak of the numerator (see figure 2.29(b)), meaning that its argument must be:

$$\frac{Nkb}{2} \sin \theta_{p \text{ neighbour}} = \frac{Nkb}{2} \sin \theta_p + \frac{3\pi}{2}$$

$$\text{and thus} \quad \frac{kb}{2} \sin \theta_{p \text{ neighbour}} = \frac{kb}{2} \sin \theta_p + \frac{3\pi}{2N}$$

From what we deduce that:

$$\sin^2 \left(\frac{Nkb}{2} \sin \theta_{p \text{ neighbour}} \right) = \cos^2 \left(\frac{Nkb}{2} \sin \theta_p \right)$$

$$\sin^2 \left(\frac{kb}{2} \sin \theta_{p \text{ neighbour}} \right) = \sin^2 \left(\frac{kb}{2} \sin \theta_p + \frac{3\pi}{2N} \right)$$

We also know that:

$$\frac{kb}{2} \sin \theta_p = p\pi$$

Thus:

$$\cos^2 \left(\frac{Nkb}{2} \sin \theta_p \right) = \cos^2(Np\pi) = ((-1)^{Np})^2 = 1$$

$$\sin^2 \left(\frac{kb}{2} \sin \theta_p + \frac{3\pi}{2N} \right) = \sin^2 \left(p\pi + \frac{3\pi}{2N} \right) = \left((-1)^p \sin \left(\frac{3\pi}{2N} \right) \right)^2 = \sin^2 \left(\frac{3\pi}{2N} \right)$$

If N is big then here again we can use expression (1.7). Then finally:

$$I_{p\,\text{neighbour}} = \left(\frac{2N}{3\pi}\right)^2$$

Finally, the ratio R is equal to:

$$R = \frac{I_{p\,\text{neighbour}}}{I_p} = \left(\frac{2N}{3\pi}\right)^2 \times \frac{1}{N^2} = \frac{4}{9\pi^2} \simeq 0.045$$

That ratio is very small compared to 1.

4. From the answer to the previous question, we know that the width of the diffraction peaks given by a grating for a single given wavelength is only equal to the width of the main peak. Let us determine the width of the peak from order p.

The central peak of order p for wavelength λ happens when:

$$\frac{kb}{2}\sin\theta_p = p\pi \qquad (2.17)$$

For these values of $\sin\theta_p$, it turns out that the numerator also cancels since:

$$\sin^2\left(\frac{kbN\sin\theta_p}{2}\right) = \sin^2(Np\pi) = 0$$

The closest angle θ_{p+} for which the numerator cancels again (which is what rules the peak width) is thus given by:

$$\frac{kbN\sin\theta_{p+}}{2} = Np\pi + \pi = \frac{Nkb\sin\theta_p}{2} + \pi$$

from which we get:

$$\sin\theta_{p+} - \sin\theta_p = \frac{2\pi}{Nkb}$$

From the definition of the derivative of a function (1.5), we know that:

$$\sin(\theta + \delta\theta) - \sin(\theta) = \frac{d\sin\theta}{d\theta}\delta\theta \text{ if } \delta\theta \text{ is small enough}$$

$$= \cos\theta\delta\theta$$

Thus in our case, if we assume that N is big enough, we can write:

$$\delta\theta = \frac{\sin\theta_{p+} - \sin\theta_p}{\cos\theta_p} = \frac{\dfrac{2\pi}{Nkb}}{\sqrt{1 - \left(p\dfrac{\lambda}{b}\right)^2}} = \frac{\lambda}{Nb\sqrt{1 - \left(p\dfrac{\lambda}{b}\right)^2}}$$

where we used that $\cos(\arcsin(u)) = \sqrt{1 - u^2}$ with $u \in \mathbb{R}$ (that formula can be retrieved from $\cos^2 u + \sin^2 u = 1$).

We can conclude from that latter result that the width of the peaks increases from an order to the next, and that the more slits the light passes through, the narrower are the peaks.

References

[1] BIPM 2018 *Proc. of the 26th Meeting of the General Conf. on Weights and Measures* p 472

[2] Schaub M *et al* 2016 *Molded Optics, Design and Manufacture* (Boca Raton, FL: CRC Press)

[3] Maurel A 2003 *Optique Ondulatoire* (Paris: Belin)

[4] Barish B C and Weiss R 1999 Ligo and the detection of gravitational waves *Phys. Today* **52** 44–50

[5] Born M and Wolf E 1999 *Principles of Optics* 7th edn (Cambridge: Cambridge University Press)

[6] Duffait R 1997 *Expériences D'optique, Agrégation de Sciences Physiques* 2nd edn (Paris: Bréal)

[7] OpenStax 2022 Diffraction https://phys.libretexts.org/@go/page/4512

[8] Reinhard E *et al* 2008 *Color Imaging: Fundamentals and Applications* (Boca Raton, FL: CRC Press)

[9] Berthier S 2003 *Iridescences, les Couleurs Physiques des Insectes* (Paris: Springer)

Hélène Ollivier and Osvaldo de Melo

Chapter 3

Fourier transform in 1D

This chapter introduces the fundamental principles of one-dimensional Fourier transform. It starts with a comprehensive introduction to the Fourier theorem for periodic functions explaining how these functions can be decomposed into their constituent harmonic components. The succession of square pulses is resolved in detail, and this serves as the basis for extending the theorem to include non-periodical functions. Then, the Fourier transform properties (scaling, translation and linearity) are demonstrated. A paragraph is devoted to the delta and comb functions, their definitions, properties, and their Fourier transforms. As useful examples, the Fourier transforms of the sine, cosine and Gaussian functions are determined. At the end of the chapter an explanation about the meaning of negative frequencies is presented. Some proposed or resolved problems are also included in the chapter.

3.1 Periodic functions. Fourier series. Examples. Succession of square pulses. Frequency spectrum

3.1.1 Periodic functions. Fourier series

When applied to a periodic function Fourier's theorem states that:

> Any periodic function can be represented by a collection (series) of harmonic functions with appropriate amplitudes, frequencies and phases.

This powerful theorem provides a method to decompose any periodic function into its harmonic components (see section 1.6). The frequencies of the harmonic functions are integer multiples of the frequency of the original function being decomposed. For periodic spatial functions, the theorem can be expressed mathematically as:

$$f(x) = \sum_{n=-\infty}^{\infty} C_n \exp\left(i2\pi n \xi_0 x\right) \tag{3.1}$$

doi:10.1088/978-0-7503-6392-1ch3

In this expression, C_n is the nth coefficient and represents the amplitude of the harmonic $\exp(i2\pi n\xi_0 x)$, where $n\xi_0$ is the spatial frequency, x is the spatial coordinate, i is the imaginary number ($i = \sqrt{-1}$) (see section 1.5), ξ_0 is the frequency of the function, and $L = 1/\xi_0$, is its spatial period. The product $2\pi n\xi_0$ is often referred to as the angular spatial frequency. The spatial period of the nth harmonic is the inverse of this spatial frequency ($1/n\xi_0$). Writing explicitly the first terms for $n \geqslant 0$:

$$f(x) = C_0 + C_1 \exp(i2\pi\xi_0 x) + C_2 \exp(i2\pi(2\xi_0)x) + C_3 \exp(i2\pi(3\xi_0)x) + \ldots \quad (3.2)$$

The coefficients C_n depend on the functional form of $f(x)$ and must be calculated. To compute these coefficients, we can use the following relation:

$$\frac{1}{L} \int_{x_0}^{x_0+L} \exp(i2\pi n\xi_0 x)[\exp(i2\pi m\xi_0 x)]^* \mathrm{d}x \equiv \frac{1}{L} \int_{x_0}^{x_0+L} \exp(i2\pi(n-m)\xi_0 x)\mathrm{d}x \quad (3.3)$$

Here, $[\exp(i2\pi m\xi_0 x)]^*$ is the complex conjugate of the harmonic of frequency $m\xi_0$. The integral in (3.3) is equal to 0 if $n \neq m$, and equal to 1 if $n = m$. This arises from the fact that, when $n \neq m$, the product of the nth harmonic ($\exp(i2\pi n\xi_0 x)$) by the complex conjugate of the mth-harmonic ($\exp(-i2\pi m\xi_0 x)$) is itself a harmonic:

$$\exp(i2\pi n\xi_0 x)\exp(-i2\pi m\xi_0 x) = \exp(i2\pi(n-m)\xi_0 x)$$

and the integral of a harmonic over an integer number of periods is zero because the positive and the negative areas defined by the function cancel out. However, when $n = m$, the exponent becomes zero and the exponential term equals 1. In this case, the integral results in L, and equation (3.3) becomes unity.

To calculate the coefficients C_n, we multiply equation (3.1) by $1/L \exp(-i2\pi m\xi_0 x)$ on both sides and integrate over a full period:

$$\int_{x_0}^{x_0+L} 1/L\, f(x)\exp(-i2\pi m\xi_0 x)\mathrm{d}x = 1/L \int_{x_0}^{x_0+L} \sum_{n=-\infty}^{\infty} C_n \exp(i2\pi n\xi_0 x)\exp(-i2\pi m\xi_0 x)\mathrm{d}x$$

Under certain conditions (for example, if $f(x)$ is continuous and the Fourier series converges) expected in physical contexts, we can exchange the sum and the integral so that the previous expression is equal to:

$$\sum_{n=-\infty}^{\infty} C_n 1/L \int_{x_0}^{x_0+L} \exp(i2\pi(n-m)\xi_0 x)\mathrm{d}x$$

According to equation (3.3), all terms in the summation vanish except for the case $n = m$, where:

$$1/L \int_{x_0}^{x_0+L} \exp(i2\pi(n-m)\xi_0 x)\mathrm{d}x = 1$$

This means that the term C_m remains. Thus, we obtain the final expression for the coefficients:

$$C_m = \int_{x_0}^{x_0+L} 1/L\, f(x)\exp(-i2\pi m\xi_0 x)\mathrm{d}x \quad (3.4)$$

which allows us to calculate the coefficients of the series.

3.1.2 The succession of square pulses

To illustrate the procedure of determining the coefficients of equation (3.1) let us consider the periodic function represented in figure 3.1(a). This function consists of a repeated rectangular shape with a base a and height 1, and a spatial period L. This function is often referred as 'the square wave', although this name is not entirely accurate since the function does not necessarily represent a wave; it can, for example, depict a static distribution of illumination. For simplicity, however, we will use this term hereafter. In optics, this function can represent a diffraction grating, where a corresponds with the width of the diffraction slits and L to the spatial period. According to the definition of spatial frequency ξ, the spatial period is related to it by $L = \frac{1}{\xi_0}$.

To calculate the coefficients of the Fourier series, we utilize equation (3.4):

$$C_n = 1/L \int_{x_0}^{x_0+L} f(x) \exp(-i2\pi n\xi_0 x)\mathrm{d}x$$

And perform the integration over one spatial period, from $-L/2$ to $L/2$:

$$C_n = 1/L \int_{-L/2}^{L/2} f(x) \exp(-i2\pi n\xi_0 x)\mathrm{d}x$$

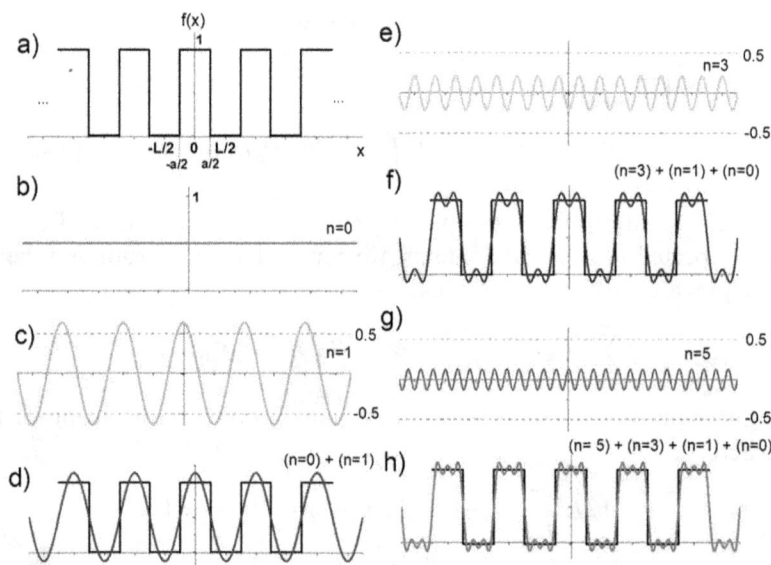

Figure 3.1. (a) Periodic succession of rectangular functions with spatial period L and non-zero region a. (b) Zero-order Fourier term ($n = 0$). (c) First-order Fourier term ($n = 1$, fundamental frequency). (d) Sum of the $n = 0$ and $n = 1$ terms, showing alongside $f(x)$. (e) Fourier term corresponding to $n = 3$. (f) Sum of the $n = 0$, $n = 1$ and $n = 3$ terms. (g) Fourier term with $n = 5$. (h) Sum of the $n = 0$, $n = 1$, $n = 3$ and $n = 5$. From (b) to (h), it was considered $L = 1$; $a = 0.5$.

We decompose this integral as follows:

$$C_n = 1/L \left[\int_{-L/2}^{-a/2} f(x) \exp(-i2\pi n\xi_0 x)dx + \int_{-a/2}^{a/2} f(x) \exp(-i2\pi n\xi_0 x)dx \right.$$
$$\left. + \int_{a/2}^{L/2} f(x) \exp(-i2\pi n\xi_0 x)dx \right]$$

Since $f(x) = 0$ in the ranges $[-L/2, -a/2]$ and $[a/2, L/2]$, the first and third integrals vanish. In the middle range, where $f(x) = 1$, the equation simplifies to:

$$C_n = 1/L \int_{-a/2}^{a/2} \exp(-i2\pi n\xi_0 x)dx \tag{3.5}$$

Then, performing the integration in equation (3.5):

$$C_n = 1/L \frac{\exp(-i2\pi n\xi_0 x)}{-i2\pi n\xi_0} \Big|_{-a/2}^{a/2} = 1/L \frac{1}{-i2\pi n\xi_0} [\exp(-i2\pi n\xi_0 a/2) - \exp(i2\pi n\xi_0 a/2)]$$

Rewriting:

$$C_n = \frac{1}{Li2\pi n\xi_0} [\exp(i2\pi n\xi_0 a/2) - \exp(-i2\pi n\xi_0 a/2)]$$

Using the Euler relation (equation (1.10)):

$$\frac{1}{2i} [\exp(i2\pi n\xi_0 a/2) - \exp(-i2\pi n\xi_0 a/2)] = \sin(\pi n\xi_0 a)$$

The coefficient C_n becomes:

$$C_n = \frac{1}{L\pi n\xi_0} \sin(\pi n\xi_0 a) \tag{3.6}$$

Multiplying by a/a:

$$C_n = \frac{a}{La\pi n\xi_0} \sin(\pi n\xi_0 a) = \frac{a}{L}\mathrm{sinc}(\pi n\xi_0 a)$$

where we have used the definition of cardinal sin or sinc function (see equation (1.13)):

$$\mathrm{sinc}(\theta) = \frac{\sin(\theta)}{\theta}$$

Knowing the coefficients C_n we can express the terms of the Fourier series using equation (3.1):

$$f(x) = \sum_{n=-\infty}^{\infty} \frac{a}{L}\mathrm{sinc}(\pi n\xi_0 a)\exp(i2\pi n\xi_0 x) \tag{3.7}$$

Let us consider a particular case with the following parameters: $\xi_0 = 1$; $a = 0.5$ and $L \equiv \frac{1}{\xi_0} = 1$. For these parameters, the Fourier series becomes:

$$f(x) = \sum_{n=-\infty}^{\infty} \frac{1}{2}\operatorname{sinc}(\frac{\pi}{2}n)\exp\,(i2\pi nx) \qquad (3.8)$$

Terms are interpreted as follows. The 0th term will be $\frac{1}{2}$ since $\lim_{\theta \to 0} \operatorname{sinc}(\theta) = 1$. This term, plotted in figure 3.1(b), is a constant equal to the mean value of the function, shifting the baseline of the other terms to $\frac{1}{2}$. The rest of the terms are harmonic functions with frequencies $\xi_0, 2\xi_0, 3\xi_0,\ldots$, or in this case ($\xi_0 = 1$), 1, 2, 3, and so on.

Using the Euler formula ($\exp\,(i\theta) = \cos\theta + i\sin\theta$) (equation (1.11)), the series can be written as:

$$f(x) = \sum_{n=-\infty}^{\infty} \frac{1}{2}\operatorname{sinc}\!\left(\frac{\pi}{2}n\right)[\cos\,(2\pi nx) + i\sin\,(2\pi nx)]$$

This sum can be divided into three parts: the 0th term; the summation for $n > 0$; and the summation for $n < 0$. So:

$$f(x) = \frac{1}{2} + \sum_{n=1}^{\infty}\frac{1}{2}\operatorname{sinc}\!\left(\frac{\pi}{2}n\right)[\cos\,(2\pi nx) + i\sin\,(2\pi nx)]$$
$$+ \sum_{n=-\infty}^{-1}\frac{1}{2}\operatorname{sinc}\!\left(\frac{\pi}{2}n\right)[\cos\,(2\pi nx) + i\sin\,(2\pi nx)]$$

$$f(x) = \frac{1}{2} + \sum_{n=1}^{\infty}\frac{1}{2}\operatorname{sinc}\!\left(\frac{\pi}{2}n\right)\cos\,(2\pi nx) + \sum_{n=1}^{\infty}\frac{1}{2}\operatorname{sinc}\!\left(\frac{\pi}{2}n\right)i\sin\,(2\pi nx) +$$
$$\sum_{n=-\infty}^{-1}\frac{1}{2}\operatorname{sinc}\!\left(\frac{\pi}{2}n\right)\cos\,(2\pi nx) + \sum_{n=-\infty}^{-1}\frac{1}{2}\operatorname{sinc}\!\left(\frac{\pi}{2}n\right)i\sin\,(2\pi nx)$$

Let us do now a simplification. Since sinc and cosine functions are even, the first and third summations are equal, and every term for positive n equals its corresponding negative term. The sine terms, being odds, cancel out, and equation (3.8) reduces to:

$$f(x) = \frac{1}{2} + 2\sum_{1}^{\infty}\frac{1}{2}\operatorname{sinc}\!\left(\frac{\pi}{2}n\right)\cos\,(2\pi nx)$$

which further simplifies to:

$$f(x) = \frac{1}{2} + \sum_{1}^{\infty}\operatorname{sinc}\!\left(\frac{\pi}{2}n\right)\cos\,(2\pi nx)$$

In this last equation it is important bearing in mind that the term for every n collects two terms: one corresponding to the positive n and one corresponding to the negative n. As the frequency of every harmonic is $n\xi_0$, this means that every term collects harmonics whose frequencies have the same values and different signs. We will deal with the meaning of negative frequencies at the end of the chapter.

Using the definition of sinc:

$$f(x) = \frac{1}{2} + \sum_{1}^{\infty} \frac{\sin\left(\frac{\pi}{2}n\right)}{\frac{\pi}{2}n} \cos(2\pi nx) = \frac{1}{2} + \sum_{1}^{\infty} \frac{2}{\pi n} \sin\left(\frac{\pi}{2}n\right) \cos(2\pi nx)$$

Let us now consider the different harmonics:

The term $n = 1$ corresponds to the harmonic with frequency ξ_0, known as the fundamental frequency. Its value is:

$$\frac{2}{\pi n} \sin\left(\frac{\pi}{2}n\right) \cos(2\pi nx) = \frac{2}{\pi} \sin\left(\frac{\pi}{2}\right) \cos(2\pi x) = \frac{2}{\pi} \cos(2\pi x)$$

This harmonic is depicted in figure 3.1(c). The sum of this harmonic and the constant term ($n = 0$) is plotted in figure 3.1(d) together with $f(x)$ to allow for comparison. The harmonic functions for even n are zero because of the factor $\sin\left(\frac{n\pi}{2}\right)$.

The harmonics with $n = 3$ and $n = 5$ are shown in figures 3.1(e) and (g) and the incremental sums are represented in figures 3.1(f) and (h). Remarkably, even with only the first few terms of the series, the sum approximates the square wave function quite well. It can be noted that high frequency harmonics are needed for accurately reproducing the regions with abrupt changes in the function.

3.1.3 The frequency spectrum

A useful and compact form to represent the Fourier series is through the so-called frequency spectrum. This representation is a graph of the amplitude of the harmonics as a function of the frequency. For the square wave, the frequency spectrum is represented in figure 3.2.

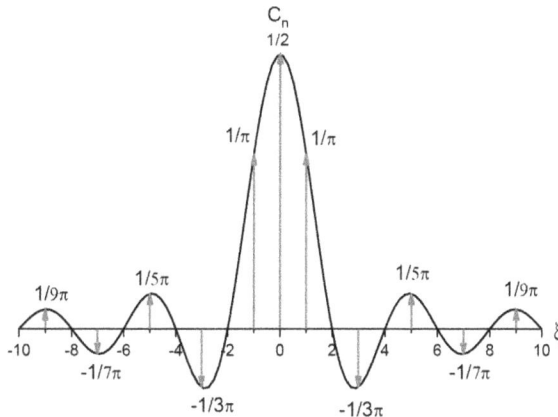

Figure 3.2. Frequency spectrum for the square wave of figure 3.1 (C_n as a function of the harmonic frequency).

3.2 Fourier transform in 1D. The square pulse

Fourier's theorem can be extended to include non-periodic functions as well. To illustrate this modification, let us consider the square wave discussed in the previous section. By allowing the spatial period of the square wave to approach infinity, the function becomes a non-periodic individual square pulse. This is because a periodic function with an infinite period is, effectively, non-periodic. Such an individual pulse is illustrated in figure 3.3.

3.2.1 From discrete to continuous

To analyse how the frequency spectrum changes as the spatial period of the square wave increases, we start from the Fourier series derived earlier (equation (3.7)):

$$f(x) = \sum_{-\infty}^{\infty} \frac{a}{L} \text{sinc}(\pi n \xi_0 a) \exp(i2\pi n \xi_0 x)$$

where the coefficients are $\frac{a}{L}\text{sinc}(\pi n \xi_0 a)$.

The first zero of the sinc function occurs when $\pi n \xi_0 a = \pi$, or equivalently, at $n\xi_0 = \frac{1}{a}$. In the previous case ($a = \frac{1}{2}$), this zero corresponded to $n\xi_0 = 2$. Since the fundamental frequency was $\xi_0 = 1$, this implies $n = 2$. Thus, within the central lobe of the sinc function, only the frequencies ξ_0 and $-\xi_0$ corresponding with $n = \pm 1$ remain.

If we now increase the spatial period to $5L$ ($\xi_0 = 1/5$), the first zero of the sinc function will still correspond with $n\xi_0 = \frac{1}{a} = 2$. However, with $\xi_0 = 1/5$, this implies $n = 10$. Consequently, the first nine frequencies ($n = 1, 2, \ldots, 9$) remain inside the central lobe. Figure 3.4 illustrates the frequency spectrum for this case, focusing on the range $[-8, 8]$ in the frequency domain. Note how the spacing between consecutive frequencies decreases from 1 to 0.2 as the spatial period increases from 1 to 5.

What happens as $L \to \infty$ ($\xi \longrightarrow 0$)? In this limit, the frequency spectrum transitions from being discrete to continuous. So, for non-periodic functions, the

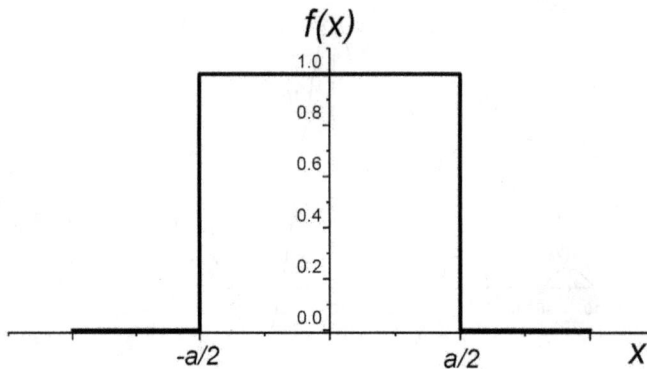

Figure 3.3. The square pulse obtained from the square wave by letting its period become infinite.

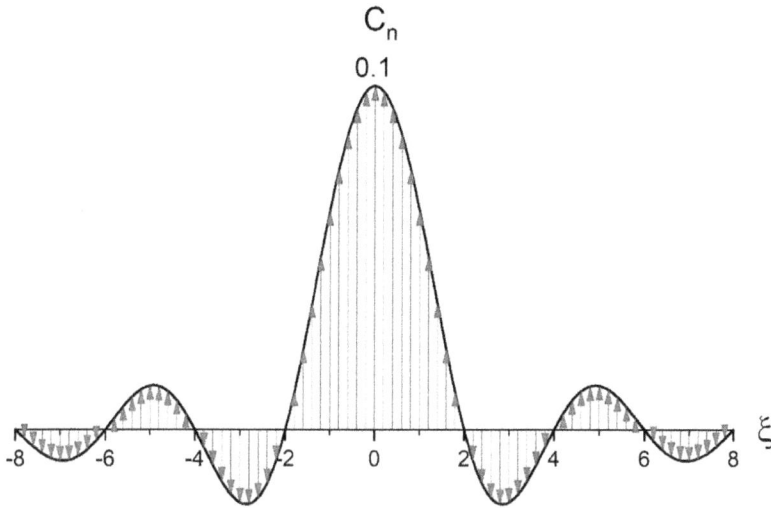

Figure 3.4. Frequency spectrum for the square wave with $\xi_0 = 0.2$ and $a = \frac{1}{2}$. Note how the spacing between frequencies decreases as the spatial period increases from $L = 1$ to $L = 5$.

frequency domain variable becomes ξ instead of $n\xi_0$, and the series is replaced by an integral:

$$f(x) = \int_{-\infty}^{\infty} F(\xi)\exp(i2\pi\xi x)d\xi \tag{3.9}$$

In this integral $F(\xi)d\xi$ plays the role of C_n. $F(\xi)$ is called the Fourier transform of the function $f(x)$. The Fourier transform is computed using:

$$F(\xi) = \int_{-\infty}^{\infty} f(x)\exp(-i2\pi\xi x)dx \tag{3.10}$$

To explicitly indicate the function being transformed, we may write $F(\xi) = F\{f(x)\}$. Equation (3.10) is called the Fourier transform, while equation (3.9) is referred to as the inverse Fourier transform. These equations allow us to switch between two domains: the spatial domain $f(x)$ and spatial frequency domain $F(\xi)$. In this context, $f(x)$ and $F(\xi)$ are referred to as a Fourier pair. Similarly, $f(t)$ and $F(\nu)$, another important Fourier pair, are frequently used in electronics to relate time and temporal frequency.

3.2.2 The square pulse

We now calculate the Fourier transform of the square pulse shown in figure 3.3. Since this function will appear frequently throughout the book, we define it as $\text{Rect}_a(x)$. This function can be analytically expressed as:

$$\text{Rect}_a(x) \equiv f(x) = \begin{cases} 1, \text{ for } |x| < a/2 \\ 0, \text{ for } |x| \geqslant a/2 \end{cases} \tag{3.11}$$

Using equation (3.10), the Fourier transform is:

$$F(\xi) = \int_{-\infty}^{\infty} \text{Rect}_a(x) \exp(-i2\pi\xi x)dx$$

This integral can be divided into three parts:

$$F(\xi) = \int_{-\infty}^{-a/2} \text{Rect}_a(x) \exp(-i2\pi\xi x)dx + \int_{-a/2}^{a/2} \text{Rect}_a(x) \exp(-i2\pi\xi x)dx$$
$$+ \int_{a/2}^{\infty} \text{Rect}_a(x) \exp(-i2\pi\xi x)dx$$

Since $\text{Rect}_a(x) = 0$ outside $[-a/2, a/2]$, The first and third integrals vanish. For the remaining integral:

$$F(\xi) = \int_{-a/2}^{a/2} \exp(-i2\pi\xi x)dx$$

Computing these integral yields:

$$F(\xi) = \left. \frac{\exp(-i2\pi\xi x)}{-i2\pi\xi} \right|_{-a/2}^{a/2} = \frac{1}{\pi\xi}\frac{1}{-2i}[\exp(-i2\pi\xi a/2) - \exp(i2\pi\xi a/2)] = \frac{1}{\pi\xi} \sin(\pi\xi a)$$

where we have made use of the Euler relation $\sin(\pi\xi a) = \frac{1}{2i}[\exp(i\pi\xi a) - \exp(-i\pi\xi a)]$. Now, multiplying by a/a and using the definition of the cardinal sine:

$$F(\xi) = \frac{a}{\pi\xi a} \sin(\pi\xi a) = a \, \text{sinc}(\pi\xi a) \tag{3.12}$$

This result resembles the expression for C_n, but here the frequency variable ξ is continuous. Figure 3.5 illustrates this Fourier transform.

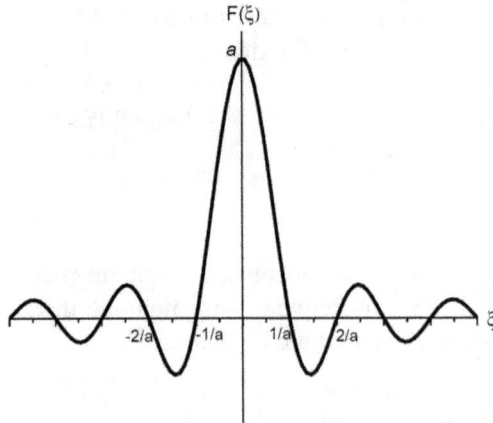

Figure 3.5. The Fourier transform of the square pulse.

Using equation (3.9), the square pulse (equation (3.11)) can be expressed as:

$$f(x) = \int_{-\infty}^{\infty} F(\xi)\exp(i2\pi\xi x)\mathrm{d}\xi = \int_{-\infty}^{\infty} a\,\mathrm{sinc}(\pi\xi a)\exp(i2\pi\xi x)\mathrm{d}\xi.$$

3.3 Properties of the Fourier transform. Linearity, scaling, translation. The Fourier transform of the Fourier transform

In the section we demonstrate three key properties of the Fourier transform: linearity, scaling, and translation. Additionally, we calculate the Fourier transform of the Fourier transform, of interest in the image formation process, and present the Fourier transform of the derivative of a function.

3.3.1 Linearity theorem

Linearity states that:

$$F\{af(x) + bg(x)\} = aF\{f(x)\} + bF\{g(x)\} \tag{3.13}$$

Here, a, b are constant (which can, in general, be complex).

The proof of this theorem is straightforward:

$$F\{af(x) + bg(x)\} = \int_{-\infty}^{\infty} [af(x) + bg(x)]\exp(-i2\pi\xi x)\mathrm{d}x$$

Since integrals are linear, we can write:

$$\int_{-\infty}^{\infty} [af(x) + bg(x)]\exp(-i2\pi\xi x)\mathrm{d}x = a\int_{-\infty}^{\infty} f(x)\exp(-i2\pi\xi x)\mathrm{d}x + b\int_{-\infty}^{\infty} g(x)\exp(-i2\pi\xi x)\,\mathrm{d}x$$
$$= aF\{f(x)\} + bF\{g(x)\}.$$

3.3.2 Theorem of translation

This theorem states that:

If $F\{f(x)\} = F(\xi)$, the Fourier transform of the translated function will be:

$$F\{f(x - x_0)\} = F(\xi)\exp(-i2\pi\xi x_0) \tag{3.14}$$

For demonstrating this theorem, consider:

$$F\{f(x - x_0)\} = \int_{-\infty}^{\infty} f(x - x_0)\exp(-i2\pi\xi x)\mathrm{d}x$$

Let $y = x - x_0$, so, $\mathrm{d}y = \mathrm{d}x$

$$F\{f(x - x_0)\} = \int_{-\infty}^{\infty} f(y)\exp(-i2\pi\xi(y + x_0))\mathrm{d}y = \int_{-\infty}^{\infty} f(y)\exp(-i2\pi\xi y)\exp(-i2\pi\xi x_0)\mathrm{d}y$$

The factor $\exp(-i2\pi\xi x_0)$ does not depend on y, so it can be factored out:

$$F\{f(x - x_0)\} = \exp(-i2\pi\xi x_0)\int_{-\infty}^{\infty} f(y)\exp(-i2\pi\xi y)\mathrm{d}y$$

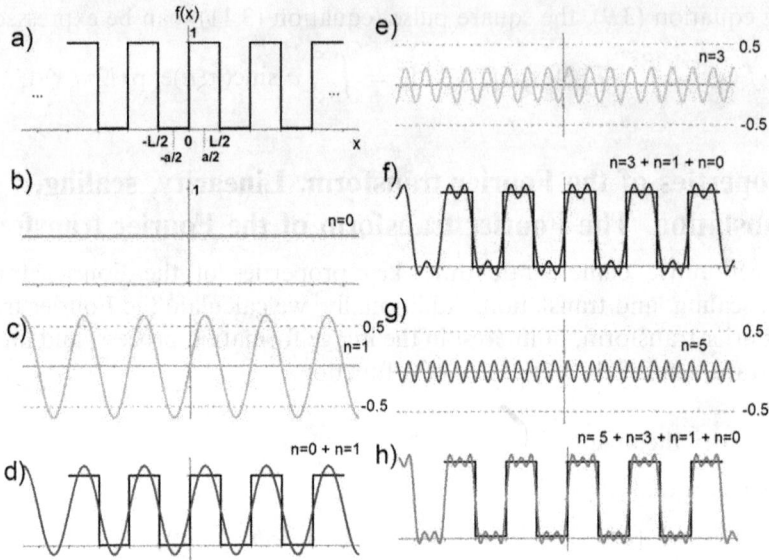

Figure 3.6. (a) Periodic succession of rectangular functions as in figure 3.1 but shifted in $\frac{a}{2}$ toward positive x. (b–h) as in figure 3.1 with the harmonic phase-shifted in $2\pi\xi\frac{a}{2}$ as well.

Now, as y is a dummy variable (the result of the integral will depend only in ξ). The integral $\int_{-\infty}^{\infty} f(y)\exp(-i2\pi\xi y)\mathrm{d}y$ represents the Fourier transform of $f(x)$. Therefore:

$$F\{f(x - x_0)\} = F(\xi)\exp(-i2\pi\xi x_0)$$

That means that the factor $\exp(-i2\pi\xi x_0)$ affects all the harmonic functions changing their phase.

For example, consider the square wave function shown in figure 3.1(a) but shifted towards the positive x values by $a/2$. To reproduce this function following equation (3.1), it is sufficient to shift the phase of all harmonic functions calculated for the un-shifted function by $2\pi\xi\frac{a}{2}$. Note that multiplying the harmonic $\exp(i2\pi\xi x)$ by $\exp(-i2\pi\xi\frac{a}{2})$ leads to $\exp(i2\pi\xi(x - \frac{a}{2}))$. The corresponding plots are shown in figure 3.6.

3.3.3 Scaling theorem

This theorem is expressed as:

$$F\{f(ax)\} = \frac{1}{|a|}F\left(\frac{\xi}{a}\right) \tag{3.15}$$

To demonstrate this, let us considerer:

$$F\{f(ax)\} = \int_{-\infty}^{\infty} f(ax)\exp(-i2\pi\xi x)\mathrm{d}x$$

Using the variable change $y = ax$, so $dy = a\,dx$ we obtain:

$$F\{f(ax)\} = \int_{-\infty}^{\infty} f(y) \exp\left(-i2\pi\xi\frac{y}{a}\right)\frac{dy}{a} = \frac{1}{a}\int_{-\infty}^{\infty} f(y) \exp\left(-i2\pi\frac{\xi}{a}y\right)dy = \frac{1}{a}F\left(\frac{\xi}{a}\right)$$

If $a > 0$ this expression will be identical to:

$$\frac{1}{|a|}F\left(\frac{\xi}{a}\right)$$

In the case $a < 0$, the change of variable implies that $y = -|a|x$ and the limits of the integral must be interchanged. That is:

$$\frac{1}{-|a|}\int_{\infty}^{-\infty} f(y) \exp\left(-i2\pi\frac{\xi}{a}y\right)dy$$

To recover the form of the Fourier transform:

$$\frac{1}{-|a|}\left[-\int_{-\infty}^{\infty} f(y) \exp\left(-i2\pi\frac{\xi}{a}y\right)dy\right] = \frac{1}{|a|}\int_{-\infty}^{\infty} f(y) \exp\left(-i2\pi\frac{\xi}{a}y\right)dy$$

It can be noted that the denominator of the integral pre-factor is always $|a|$ regardless the sign of a.

Qualitatively, this theorem means that if $a > 1$ (the function 'shrinks' in x), the Fourier transform will 'stretch' in ξ, and vice versa. The reader can verify that this theorem is satisfied for the square pulse, for example.

3.3.4 The Fourier transform of the Fourier transform

Applying the inverse Fourier transform to the Fourier transform returns the original function. What happens if we apply the Fourier transform (not the inverse) to the Fourier transform?

$$F\{F(\xi)\} = \int_{-\infty}^{\infty} F(\xi) \exp(-i2\pi\xi x)d\xi = \int_{-\infty}^{\infty} F(\xi) \exp(i2\pi\xi(-x))d\xi$$

However, $\int_{-\infty}^{\infty} F(\xi) \exp(i2\pi\xi x)d\xi = f(x)$ according to (3.9). Then:

$$\int_{-\infty}^{\infty} F(\xi) \exp(i2\pi\xi(-x))d\xi = f(-x)$$

Or:

$$F\{F(\xi)\} = f(-x)$$

That is, the Fourier transform of the Fourier transform return the function $f(x)$ with the sign of its argument changed. If the function is even, the result is the same function. This result will be used in chapter 7 to explain image formation as the Fourier transform of the Fourier transform.

3.3.5 The Fourier transform of the derivative of a function

Let us calculate the Fourier transform of the derivative of a function, that is:

$$F\{f'(x)\} = \int_{-\infty}^{\infty} f'(x)\exp(-i2\pi\xi x)\mathrm{d}x$$

This integral can be resolved by integration by parts. Recall that $\int u\mathrm{d}v = uv - \int v\mathrm{d}u$. In this case: $u = \exp(-i2\pi\xi x)$, so $\mathrm{d}u = -i2\pi\xi \exp(-i2\pi\xi x)$; $\mathrm{d}v = f'(x)\mathrm{d}x$, so $v = f(x)$. Thus:

$$\int_{-\infty}^{\infty} f'(x)\exp(-i2\pi\xi x)\mathrm{d}x = \exp(-i2\pi\xi x)\,f(x)|_{-\infty}^{\infty} + (i2\pi\xi)\int f(x)\exp(-i2\pi\xi x)\mathrm{d}x$$

If the function vanishes at infinity, the first term in the right side of the equation is 0 and:

$$F\{f'(x)\} = i2\pi\xi\, F\{f(x)\}$$

The Fourier transform of higher order derivatives can be calculated using the recurrence relation:

$$F\{f''(x)\} = i2\pi\xi\, F\{f'(x)\} = (i2\pi\xi)^2 F\{f(x)\}$$

And, in general:

$$F\{f^{n}(x)\} = (i2\pi\xi)^n F\{f(x)\}$$

This result will be used in the problem 3.6.

3.4 The Dirac delta function

3.4.1 The Dirac delta function. Properties. The sifting property

In simple terms, the Dirac delta function (hereafter referred as delta function) can be seen as an infinitely tall and infinitely narrow function. For example, forces involved in collisions can be modelled as delta functions in time because their duration is extremely short, and their magnitudes are relatively large. In optics, a very thin illuminated slit can be approximated as a Dirac delta function in space in the direction normal to the slit. Mathematically, the delta function is represented as $\delta(x)$.

More precisely, the delta function $\delta(x)$ can be conceptualized as a square pulse with an infinite height and infinitesimal width, such that the area under the curve (the integral of the function) is unity. This means:

$$\int_{-\infty}^{\infty} \delta(x)\mathrm{d}x = 1 \tag{3.16}$$

The delta function is centred at the point where its argument is zero. Thus, $\delta(x - x_0)$ represents the delta centred at x_0. In the function $A\delta(x)$, A corresponds with the area as given by equation (3.16). Delta functions are always depicted as arrows, with the height of the arrow corresponding to the area; some examples are shown in figure 3.7.

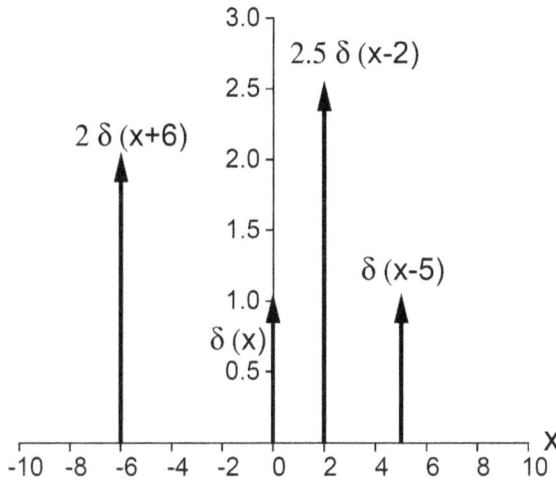

Figure 3.7. Some examples of delta functions.

The infinitesimal width of the delta function allows it to 'sift out' a single value of a function when multiplied and integrated. For example:

$$\int_{-\infty}^{\infty} f(x)\delta(x)\mathrm{d}x = f(0) \tag{3.17}$$

This property is known as the sifting property. It arises because, within the infinitesimal range of the delta function, $f(x) \approx f(0)$. Thus:

$$\int_{-\infty}^{\infty} f(0)\delta(x)\mathrm{d}x = f(0)\int_{-\infty}^{\infty}\delta(x)\mathrm{d}x = f(0)$$

Similarly:

$$\int_{-\infty}^{\infty} f(x)\delta(x - x_0)\mathrm{d}x = f(x_0) \tag{3.18}$$

Another way to interpret this property is to treat x_0 as a variable, say x. Then:

$$\int_{-\infty}^{\infty} f(x')\delta(x' - x)\mathrm{d}x' = f(x) \tag{3.19}$$

This representation will be useful when studying convolutions in chapter 5. The delta function is an even function, meaning $\delta(x) = \delta(-x)$.

3.4.2 The comb function

A periodic distribution of deltas from $-\infty$ to ∞ is called a comb function. Such a function with a spatial period $L = 2$ (spatial frequency $\xi_0 = 0.5$) is represented in figure 3.8.

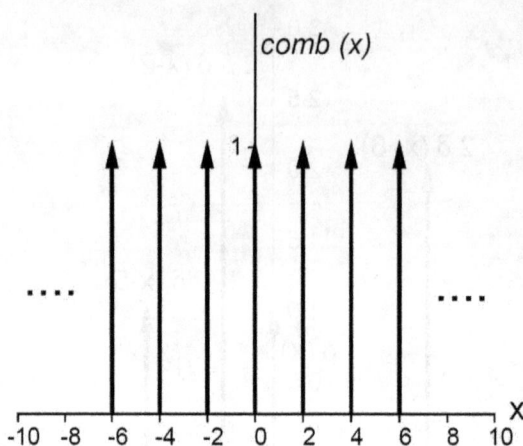

Figure 3.8. The comb function.

The comb function can be expressed analytically as:

$$\sum_{-\infty}^{\infty} \delta(x - nL) \tag{3.20}$$

with $n = \ldots -3, \; -2, \; -1, \; 0, \; 1, \; 2, \; 3, \; \ldots$.

3.4.3 The Fourier transform of delta and comb functions

The Fourier transform of the delta function will be:

$$F\{\delta(x)\} = \int_{-\infty}^{\infty} \delta(x) \exp(-i2\pi\xi x)\mathrm{d}x$$

Using the sifting property of the delta function (equation (3.17)):

$$F\{\delta(x)\} = \exp(-i2\pi\xi(0)) = 1$$

To illustrate this result, consider the scaling property of the Fourier transform (equation (3.15)). If the delta function is viewed as a square pulse with infinitesimal width and infinite height (a square pulse shrunk to zero), its Fourier transform corresponds to a sinc function expanded to infinity. While this is not rigorous, it provides an intuitive understanding of the scaling property.

The Fourier transform of the delta shifted from zero, $\delta(x - x_0)$:

$$F\{\delta(x - x_0)\} = \exp(-i2\pi\xi x_0)$$

This result can be obtained directly from the definition of the Fourier transform and the sifting property (equation (3.18)) or by applying the translation property (equation (3.14)). Since the Fourier transform of $\delta(x)$ is 1, the Fourier transform of the shifted delta function $\delta(x - x_0)$ will be $\exp(-i2\pi\xi x_0)$.

Note that the Fourier transform of the delta function can be expressed as a complex number $\rho \exp(-i\theta)$, where the modulus $\rho = 1$. For $\delta(x)$, the phase $\theta = 0$, and in the case of $\delta(x - x_0)$ the phase $\theta = -2\pi\xi x_0$.

A useful way to express the delta function which will be used later, is:

$$\delta(x) = \int_{-\infty}^{\infty} \exp(i2\pi\xi x)d\xi \qquad (3.21)$$

The integral is zero everywhere except at $x = 0$ (since the integral of a harmonic function over symmetric limits is zero) but for $x = 0$ it becomes:

$$\int_{-\infty}^{\infty} d\xi \rightarrow \infty$$

Thus, the function is zero everywhere except at $x = 0$, where it is infinite, mimicking the behaviour of the delta function.

Similarly:

$$\delta(\xi) = \int_{-\infty}^{\infty} \exp(i2\pi\xi x)dx \qquad (3.22)$$

To calculate the Fourier transform of the comb function, we use its periodicity and express it as a sum of harmonic components:

$$\text{comb}(x) = \sum_{n=-\infty}^{\infty} \delta(x - nL)$$

with $n = \ldots -3, -2, -1, 0, 1, 2, 3, \ldots$.

Alternatively, using Fourier's theorem (equation (3.1)):

$$\text{comb}(x) = \sum_{n=-\infty}^{\infty} C_n \exp(i2\pi n\xi_0 x) \qquad (3.23)$$

Here, $\xi_0 = 1/L$ is the spatial frequency of the comb function. The coefficients C_n. can be calculated using equation (3.4):

$$C_n = 1/L \int_{x}^{x+L} \text{comb}(x) \exp(-i2\pi n\xi_0 x)dx = 1/L \int_{x}^{x+L} \left[\sum_{-\infty}^{\infty} \delta(x - nL)\right] \exp(-i2\pi n\xi_0 x)dx$$

Let us consider the interval $[-L/2, L/2]$, where the comb function contains only the term $n = 0$ ($\delta(x)$). Thus:

$$C_n = 1/L \int_{-L/2}^{L/2} \delta(x)\exp(-i2\pi n\xi_0 x)dx = 1/L \exp(-i2\pi n\xi_0(0)) = 1/L$$

This means the coefficients C_n are independent of ξ and all have the same value $1/L$. The series (equation (3.23)) then becomes:

$$\text{comb}(x) = 1/L \sum_{-\infty}^{\infty} \exp(i2\pi n\xi_0 x)$$

The Fourier transform is:

$$F\{\text{comb}(x)\} = \int_{-\infty}^{\infty} 1/L \sum_{-\infty}^{\infty} \exp(i2\pi n\xi_0 x) \exp(-i2\pi\xi x)dx$$

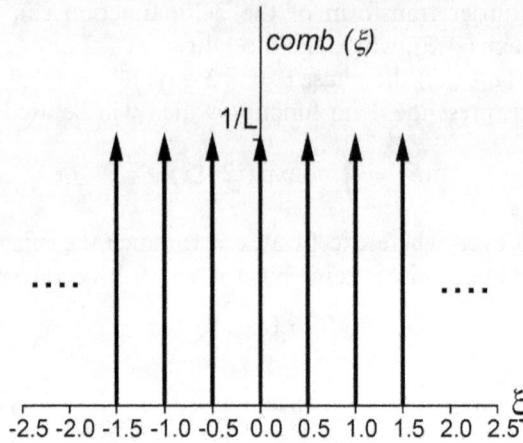

Figure 3.9. The Fourier transform of the comb function.

Interchanging the integral and summation:

$$F\{\text{comb}(x)\} = 1/L \sum_{-\infty}^{\infty} \int_{-\infty}^{\infty} \exp{(i2\pi n\xi_0 x)} \exp{(-i2\pi\xi x)} \mathrm{d}x$$

$$= 1/L \sum_{-\infty}^{\infty} \int_{-\infty}^{\infty} \exp{(i2\pi(n\xi_0 - \xi)x)} \mathrm{d}x$$

Using equation (3.22):

$$\int_{-\infty}^{\infty} \exp{(i2\pi(n\xi_0 - \xi)x)} \mathrm{d}x = \delta(n\xi_0 - \xi) = \delta(\xi - n\xi_0)$$

(we have used the fact that $\delta(x)$ is an even function).

Thus:

$$F\{\text{comb}(x)\} = 1/L \sum_{-\infty}^{\infty} \delta(\xi - n\xi_0) \tag{3.24}$$

This shows that the Fourier Transform of the comb function is also a comb function in the ξ domain. Figure 3.9 illustrate the Fourier transform of the comb function of figure 3.8 (with $\xi_0 = 0.5$, $L = 2$).

3.5 The Fourier transform of the sine, cosine, and Gaussian functions

3.5.1 The Fourier transform of the sine and cosine functions

To calculate the Fourier transform of the sine and cosine functions, we will express them using the Euler formulae (see equations (1.10) and (1.11)):

$$\sin{(\theta)} = \frac{1}{2i}(\exp{(i\theta)} - \exp{(-i\theta)}) \text{ and } \cos{(\theta)} = \frac{1}{2}(\exp{(i\theta)} + \exp{(-i\theta)})$$

Let us compute the Fourier transform of $\cos(2\pi\xi_0 x)$, using the definition of the Fourier transform and the Euler relation for the cosine function:

$$F\{\cos(2\pi\xi_0 x)\} = \int_{-\infty}^{\infty} \cos(2\pi\xi_0 x)\exp(-i2\pi\xi x)dx$$

$$= \int_{-\infty}^{\infty} \frac{1}{2}(\exp(i2\pi\xi_0 x) + \exp(-i2\pi\xi_0 x))\exp(-i2\pi\xi x)dx$$

Simplifying the integral:

$$= \int_{-\infty}^{\infty} \frac{1}{2}(\exp(i2\pi(\xi_0 - \xi)x) + \exp(-i2\pi(\xi_0 + \xi)x))dx$$

Using equation (3.21) for the terms inside the parenthesis:

$$F\{\cos(2\pi\xi_0 x)\} = \frac{1}{2}[\delta(\xi - \xi_0) + \delta(\xi + \xi_0)] \qquad (3.25)$$

Note that the Fourier transform of $\cos(2\pi\xi_0 x)$ is a real function. Its frequency spectrum is illustrated in figure 3.10.

Similarly, the Fourier transform of the $\sin(2\pi\xi_0 x)$ function can be calculated as follows:

$$F\{\sin(2\pi\xi_0 x)\} = \int_{-\infty}^{\infty} \sin(2\pi\xi_0 x)\exp(-i2\pi\xi x)dx$$

$$= \int_{-\infty}^{\infty} \frac{1}{2i}(\exp(i2\pi\xi_0 x) - \exp(-i2\pi\xi_0 x))\exp(-i2\pi\xi x)dx$$

Simplifying the integral:

$$F\{\sin(2\pi\xi_0 x)\} = \int_{-\infty}^{\infty} \frac{1}{2i}(\exp(i2\pi(\xi_0 - \xi)x) - \exp(-i2\pi(\xi_0 + \xi)x))dx$$

Using again equation (3.21):

$$F\{\sin(2\pi\xi_0 x)\} = \frac{1}{2i}\int_{-\infty}^{\infty} (\exp(i2\pi(\xi_0 - \xi)x) - \exp(-i2\pi(\xi_0 + \xi)x))dx$$

$$= -\frac{i}{2}[\delta(\xi - \xi_0) - \delta(\xi + \xi_0)]$$

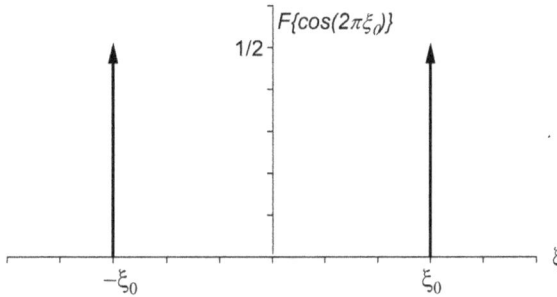

Figure 3.10. The frequency spectrum of the $\cos(2\pi\xi_0 x)$.

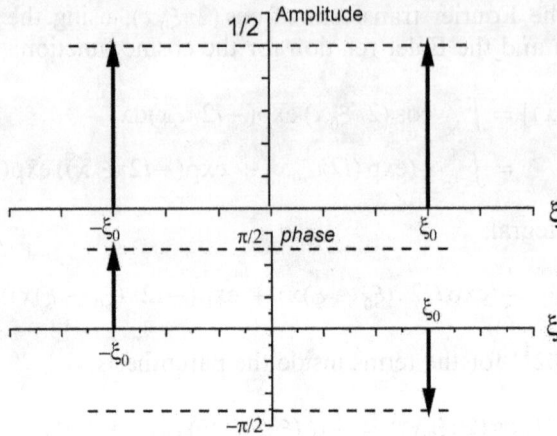

Figure 3.11. The frequency spectrum of the sin $(2\pi\xi_0 x)$.

Unlike the cosine function, the Fourier transform of $\sin(2\pi\xi_0 x)$ is a purely imaginary function. Using the Euler formula ($\exp(i\theta) = \cos\theta + i\sin\theta$, see equation (1.11)), we can substitute $\exp(i\pi/2) = i$ and $\exp(i\pi) = -1$ into the equation to obtain:

$$F\{\sin(2\pi\xi_0 x)\} = \int_{-\infty}^{\infty} \frac{1}{2}\exp(-i\pi/2)[\exp(i2\pi(\xi_0 - \xi)x) + \exp(i\pi)\exp(-i2\pi(\xi_0 + \xi)x)]dx$$

Simplifying further:

$$= \frac{1}{2}[\delta(\xi - \xi_0)\exp(-i\pi/2) + \delta(\xi + \xi_0)\exp(i\pi/2)] \tag{3.26}$$

Each term of this Fourier transform is a complex number of the form $\rho\exp(i\theta)$, where ρ is the amplitude and θ is the phase. To represent the frequency spectrum of this function, it is common to plot separate graphics for the amplitudes and phases at each frequency, as is shown in figure 3.11.

3.5.2 The Fourier transform of the Gaussian function

The Gaussian function is significant in optics, as it describes, for example, the radial profile of a laser beam intensity. It is also used as a filter (the Gaussian filter) to reduce the noise or smooth data. Its definition and graphic representation are provided in section 1.7.3. If the maximum is located at $\mu = 0$, the normalized function ($\frac{1}{\sigma\sqrt{2\pi}} = 1$) can be expressed as:

$$f(x) = \exp(-\pi x^2) \tag{3.27}$$

It is also known that:

$$\int_{-\infty}^{\infty} \exp(-x^2)dx = \sqrt{\pi} \tag{3.28}$$

Let us calculate the Fourier transform of the function in equation (3.27):

$$F(\xi) = \int_{-\infty}^{\infty} \exp(-\pi x^2)\exp(-i2\pi\xi x)dx = \int_{-\infty}^{\infty} \exp(-(\pi x^2 + i2\pi\xi x))dx \tag{3.29}$$

To evaluate this integral, we complete the square in the exponent. First, perform the change of variable:

$$u = \sqrt{\pi}x$$

and

$$du = \sqrt{\pi}dx$$

After this coordinate change, equation (3.29) becomes:

$$F(\xi) = \frac{1}{\sqrt{\pi}} \int_{-\infty}^{\infty} \exp(-(u^2 + i2\sqrt{\pi}\xi u))du$$

To complete the square add and subtract $(i\sqrt{\pi}\xi)^2$ in the exponent. Note that $u^2 + i2\sqrt{\pi}\xi u + (i\sqrt{\pi}\xi)^2$ is of the form of a perfect square $(u + b)^2 = u^2 + 2bu + b^2$ with $b = i\sqrt{\pi}\xi$. Then:

$$F(\xi) = \frac{1}{\sqrt{\pi}} \int_{-\infty}^{\infty} \exp(-(u^2 + i2\sqrt{\pi}\xi u + (i\sqrt{\pi}\xi)^2))\exp((i\sqrt{\pi}\xi)^2))du$$

This simplifies to:

$$F(\xi) = \frac{1}{\sqrt{\pi}} \int_{-\infty}^{\infty} \exp(-(u + i\sqrt{\pi}\xi)^2)\exp((i\sqrt{\pi}\xi)^2)du$$

Let $y = u + i\sqrt{\pi}\xi$, so $dy = du$. Then:

$$F(\xi) = \frac{1}{\sqrt{\pi}} \int_{-\infty}^{\infty} \exp(-(y)^2)\exp((i\sqrt{\pi}\xi)^2)dy = \frac{1}{\sqrt{\pi}} \exp((i\sqrt{\pi}\xi)^2) \int_{-\infty}^{\infty} \exp(-(y)^2)dy$$

Since $\int_{-\infty}^{\infty} \exp(-(y)^2)dy = \sqrt{\pi}$ we obtain:

$$F(\xi) = \exp(-\pi\xi^2)$$

the Fourier transform of the Gaussian function is another Gaussian function in the frequency domain. The Gaussian function and its Fourier transform are shown in figure 3.12.

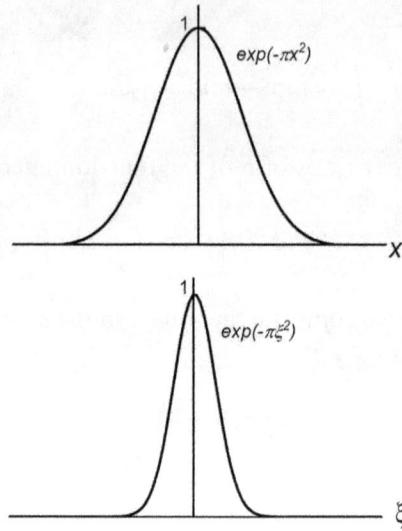

Figure 3.12. The Gaussian function (top) and its Fourier transform (bottom).

3.6 The negative frequencies

The reader may have noticed that, in this chapter, the frequency spectra of the studied functions cover both negative and positive ranges. For example, the Fourier transform of the cosine and sine functions has both one positive and one negative frequency. Shouldn't the frequency spectrum of a harmonic function such as the sine or the cosine functions be composed of a single positive frequency? Typically, the frequency of a harmonic is defined as the number of periods occurring in the unit time or length. So, what do negative frequencies signify?

Let us consider the harmonic $\cos(2\pi\xi_0 x)$. Using the complex representation:

$$\cos(2\pi\xi_0 x) = \frac{1}{2}(\exp(i2\pi\xi_0 x) + \exp(-i2\pi\xi_0 x))$$

The cosine function is half of the sum of two complex functions with phases of opposite signs. These opposite phases signs can be interpreted as opposite frequency signs:

$$\cos(2\pi\xi_0 x) = \frac{1}{2}(\exp(i2\pi\xi_0 x) + \exp(i2\pi(-\xi_0)x))$$

In other words, representing a harmonic function in complex notation, it is required to introduce both a positive and a negative frequency. While the complex representation simplifies calculations, it also introduces the concept of negative frequencies which may seem counterintuitive.

Using a graphical representation of the complex function, $\rho \exp(i2\pi\xi_0 x)$, the real and imaginary part of the cosine function can be visualized, as a function of x. The sum of the two complex exponential functions that form the cosine function are illustrated in figure 3.13(a) for the case $x = 0$. At $x = 0$, the imaginary part of both

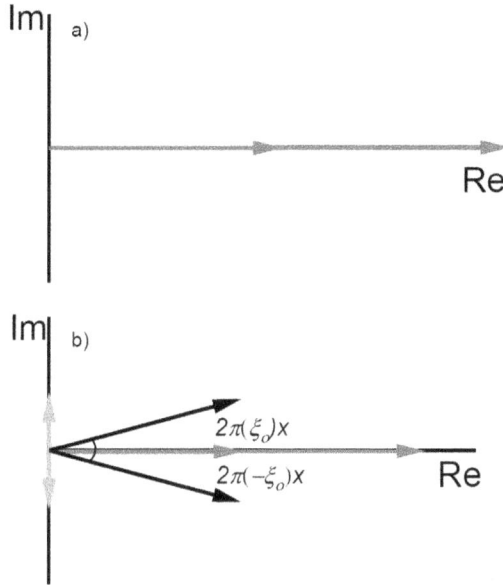

Figure 3.13. (a) Graphical representation of the complex function. (b) At $x = 0$ the imaginary parts of both exponential functions cancel each other out (b); for any value of value of x, the imaginary part (in green) cancels out.

exponentials is zero and their real components reach their maximum values of 1. As x increases, such that the phase lies between 0 and $\frac{\pi}{2}$ (see figure 3.13(b)), the term corresponding to the exponential with $\xi > 0$ rotates counterclockwise, while the one corresponding to the exponential with $\xi < 0$ will rotates clockwise. Note that the imaginary part of both complex exponentials cancels each other out for all values of x. Thus, the result of the sum of the two exponentials is simply the sum of their real parts, that is:

$$\frac{1}{2} \cos (2\pi\xi_0 x) + \frac{1}{2} \cos (2\pi\xi_0 x) = \cos (2\pi\xi_0 x)$$

In this way, the cosine function emerges from the sum of two complex exponential with phases with opposite signs (interpreted as opposite frequencies) and half amplitudes.

For the sine function, $\sin (2\pi\xi_0 x)$ we have:

$$\sin (2\pi\xi_0 x) = \frac{1}{2i}(\exp (i2\pi\xi_0 x) - \exp (i2\pi(-\xi_0)x))$$

Here, we have also two exponentials with frequencies of opposite signs. Unlike the cosine function, the sine function includes a factor of i in the denominator and a minus before the second exponential.

Using the relations $\exp(i\pi/2) = i$ and $\exp(i\pi) = -1$ we can rewrite the expression as:

$$\sin(2\pi\xi_0 x) = \frac{1}{2}[\exp(-i\pi/2)\exp(i2\pi\xi_0 x) + \exp(i\pi)\exp(-i\pi/2)\exp(i2\pi(-\xi_0)x)]$$

$$= \frac{1}{2}[\exp(-i\pi/2)\exp(i2\pi\xi_0 x) + \exp(i\pi/2)\exp(i2\pi(-\xi_0)x)] \tag{3.30}$$

$$= \frac{1}{2}\left[\exp(i(2\pi\xi_0 x - \pi/2)) + \exp(i(2\pi(-\xi_0)x + \pi/2))\right]$$

The graphical representation of this function is shown in figure 3.14(a) for $x = 0$. The real components are zero, and the imaginary part of both exponentials cancel each other out. As x increases, (see figure 3.14(b)), the modulus of the exponential with $\xi > 0$ rotates counterclockwise, starting from $-\pi/2$ while that of the exponential with $\xi < 0$ rotates clockwise, starting from $\pi/2$. The imaginary part of both complex exponential functions cancels each other out for all x values. Thus, the sum of the two exponentials is the sum of their real parts, that is:

$$\frac{1}{2}\cos(2\pi\xi_0 x - \pi/2) + \frac{1}{2}\cos(-2\pi\xi_0 x + \pi/2)$$

But the cosine function is even, which simplifies to:

$$\frac{1}{2}\cos(2\pi\xi_0 x - \pi/2) + \frac{1}{2}\cos(2\pi\xi_0 x - \pi/2) = \sin(2\pi\xi_0 x)$$

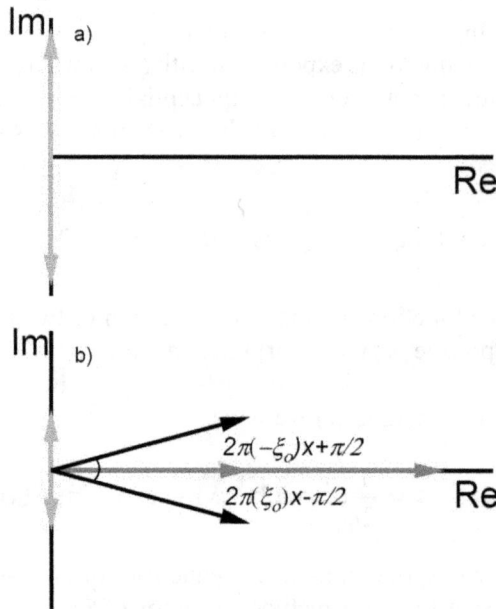

Figure 3.14. (a) At $x = 0$, the imaginary part of both exponential functions composing the sin function cancel out; (b) for any value of value of x, the imaginary part (in green) cancels out.

3.7 Problems

Problem 3.1

Demonstrate that:

$$\int_{-\infty}^{\infty} a\,\mathrm{sinc}(\pi\xi a)\exp(i2\pi\xi x)d\xi = 2\int_{0}^{\infty} a\,\mathrm{sinc}(\pi\xi a)\cos(2\pi\xi x)d\xi$$

Problem 3.2 Resolve numerically $f(x) = 2\int_{0}^{\infty} a\,\mathrm{sinc}(\pi\xi a)\cos(2\pi\xi x)d\xi$ and plot $f(x)$ for various approximations of the upper limit (e.g. 100, 1000 etc).

Problem 3.3 (with solution)

Calculate the Fourier transform for the triangular pulse represented in figure 3.15.
Solution:
The given function is:

$$f(x) = 1 + \frac{2x}{L} \text{ in the range } -\frac{L}{2} < x < 0$$

$$f(x) = 1 - \frac{2x}{L} \text{ in the range } 0 < x < \frac{L}{2}$$

The Fourier transform:

$$f(x) = \int_{-\infty}^{\infty} f(x)\exp(-2\pi i\xi x)dx$$

$$= \int_{-\frac{L}{2}}^{0}\left(1 + \frac{2x}{L}\right)\exp(-2\pi i\xi x)dx + \int_{0}^{\frac{L}{2}}(1 - \frac{2x}{L})\exp(-2\pi i\xi x)dx$$

$$= \int_{-\frac{L}{2}}^{0}\exp(-i2\pi\xi x)dx + \frac{2}{L}\int_{-\frac{L}{2}}^{0}x\,\exp(-2\pi i\xi x)dx$$

$$+ \int_{0}^{\frac{L}{2}}\exp(-i2\pi\xi x)dx - \frac{2}{L}\int_{0}^{\frac{L}{2}}x\,\exp(-i2\pi\xi x)dx$$

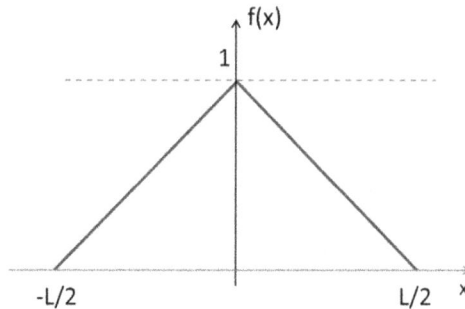

Figure 3.15. The triangular pulse.

The first and third integral can be unified in one:

$$= \int_{-\frac{L}{2}}^{\frac{L}{2}} \exp(-i2\pi\xi x)dx + \frac{2}{L}\left[\int_{-\frac{L}{2}}^{0} x \exp(-i2\pi\xi x)dx - \int_{0}^{\frac{L}{2}} x \exp(-i2\pi\xi x)dx\right]$$

The integral inside the square bracket will be resolved by integration by parts. Some algebraic steps lead to: $F\{f(x)\} = \frac{L}{2}\mathrm{sinc}^2(\pi\xi\frac{L}{2})$.

Problem 3.4 Resolve:

(a) $\int_{-\infty}^{\infty} 2x^2\delta(x)dx$ (b) $\int_{-\infty}^{\infty} 3xy\delta(x)\,dx$ (c) $\int_{-\infty}^{\infty} 3x^2\delta(x-2)dx$

Problem 3.5 Compare the Fourier transform of the succession of square pulses and the comb(x) function. What is similar and what is different in their frequency spectrum? What parameter must be changed to convert the succession of square pulses into the comb(x) function? How can the Fourier transform comb(ξ) be reached by the change of the parameter?

Problem 3.6 (with solution)

Calculate the Fourier transform of the triangular function (figure 3.15) using successive derivative over the function and applying the concepts described in section 3.3.5.

Solution:

The method involves differentiating the function iteratively until it is reduced to delta functions. As illustrated in figure 3.16, this process starts with the triangular function, followed by its first and second derivatives. By the second derivative, only delta functions persist. Discontinuities in the function are represented by delta functions, scaled by the magnitude of the function's step. The second derivative can be expressed as:

$$\frac{\delta^2 f}{\delta x^2} = \frac{2}{L}\delta\left(x + \frac{L}{2}\right) + \frac{2}{L}\delta\left(x - \frac{L}{2}\right) - \frac{4}{L}\delta(x)$$

$$\frac{\delta^2 f}{\delta x^2} = \frac{2}{L}\left[\delta\left(x + \frac{L}{2}\right) + \delta\left(x - \frac{L}{2}\right) - 2\delta(x)\right]$$

The Fourier Transform will be:

$$F\left\{\frac{\delta^2 f}{\delta x^2}\right\} = \frac{2}{L}\left(\exp\left(2\pi\xi\frac{L}{2}\right) + \exp\left(-i2\pi\xi\frac{L}{2}\right) - 2\right)$$

$$F\left\{\frac{\delta^2 f}{\delta x^2}\right\} = \frac{2}{L}(2\cos(\pi\xi L) - 2) = \left(-\frac{4}{L}(1 - \cos(\pi\xi L))\right)$$

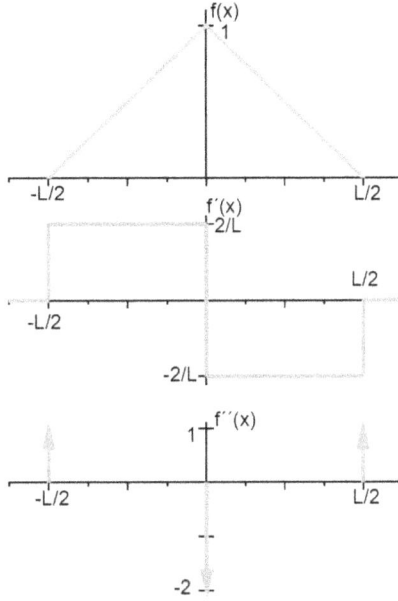

Figure 3.16. The triangular function and its first and second derivatives.

We can transform the above expression using the identity: $\sin^2(\theta) = \frac{1}{2}(1 - \cos(2\theta))$. Then:

$$F\left\{\frac{\delta^2 f}{\delta x^2}\right\} = -\frac{8}{L}\sin^2\left(\pi\xi\frac{L}{2}\right)$$

As stated above (see section 3.3.5)

$$F\left\{\frac{\delta^2 f}{\delta x^2}\right\} = (i2\pi\xi)^2 F(\xi)$$

Then:

$$(i2\pi\xi)^2 F(\xi) = -8/L \sin\left(\pi\xi\frac{L}{2}\right)\sin\left(\pi\xi\frac{L}{2}\right)$$

$$F(\xi) = \frac{L}{2}\left[\text{sinc}\left(\pi\xi\frac{L}{2}\right)\right]^2$$

which is in agreement with the result of problem 3.3.

Problem 3.7 Determine the Fourier Transform of the $\cos{(2\pi\xi_0 x)}$ limited between $-\frac{L}{2}$ and $\frac{L}{2}$.

Problem 3.8 Calculate de Fourier Transform of: $f(x) = 1 + \cos{(2\pi\xi_0 x)}$.

Problem 3.9 (with solution)

Calculate the Fourier transform of $f(x) = \cos^2(2\pi\xi_0 x)$ limited between $-L$ and L. (with solution). Represent the frequency spectrum.

$$\text{Use: } \cos^2\theta = \frac{1}{2} + \frac{1}{2}\cos 2\theta$$

$$\cos^2(2\pi\xi_0 x) = \frac{1}{2} + \frac{1}{2}\cos[2\pi(2\xi_0)x] = \frac{1}{2} + \frac{(\exp(i2\pi 2\xi_0 x) + \exp(-i2\pi 2\xi_0 x))}{4}$$

$$F(\xi) = \frac{1}{2}\int_{-L}^{L}\exp(i2\pi\xi x)dx + \frac{1}{4}\int_{-L}^{L}\exp(i2\pi(2\xi_0 - \xi)x)dx + \frac{1}{4}\int_{-L}^{L}\exp(-i2\pi(2\xi_0 + \xi)x)dx$$

Then:

$$F(\xi) = \frac{1}{2}\left(\frac{\exp(i2\pi\xi x)}{i2\pi\xi}\right)\Bigg|_{-L}^{L} + \frac{1}{4}\left(\frac{\exp(i2\pi(2\xi_0 - \xi)x)}{i2\pi(2\xi_0 - \xi)}\right)\Bigg|_{-L}^{L} + \frac{1}{4}\left(\frac{\exp(-i2\pi(2\xi_0 + \xi)x)}{-i2\pi(2\xi_0 + \xi)}\right)\Bigg|_{-L}^{L}$$

$$F(\xi) = L\mathrm{sinc}(2\pi\xi L) + \frac{L}{2}\mathrm{sinc}[2\pi(\xi - 2\xi_0)L] + \frac{L}{2}\mathrm{sinc}[2\pi(\xi + 2\xi_0)L]$$

Further Reading

[1] Bracewell R N 2000 *The Fourier Transform and Its Applications* 3rd edn (New York: McGraw-Hill)

[2] Stein E M and Shakarchi R 2003 *Fourier Analysis: An Introduction* (Princeton, NJ: Princeton University Press)

[3] James J F 2011 *A Student's Guide to Fourier Transforms* 3rd edn (Cambridge: Cambridge University Press)

[4] Beerends R J et al 2003 *Fourier and Laplace Transforms* (Cambridge: Cambridge University Press)

[5] Hecht E 2017 *Optics* 5th edn (Boston, MA: Pearson) ch 11

IOP Publishing

Elementary Fourier Optics for Science and Engineering Students

Hélène Ollivier and Osvaldo de Melo

Chapter 4

Fourier transform in 2D

This chapter introduces the 2D Fourier transform as a natural extension of the 1D Fourier transform. The meaning of the 2D harmonic components is explained through illustrative examples and visualizations. The Fourier transform of the 2D rectangular function is calculated as an example of a separable variables function. The chapter then explores a practical application: image filtering in two dimensions. By modifying the Fourier transform of an image, specific features can be enhanced or suppressed. The open-source software Image J is used to demonstrate low-pass and high-pass filtering, providing a hands-on understanding of frequency-domain manipulation. Finally, the chapter introduces the Fourier transform in polar coordinates also known as the Fourier–Bessel function. As an illustration, the Fourier transform of a cylindrical function (in optics, circular aperture) is calculated both analytically and using Image J. This highlights the utility of the Fourier–Bessel transform in problems with radial symmetry.

4.1 Fourier transform in 2D

1D Fourier transforms are widely used for time dependent functions and in some problems in optics. For example, a very large and uniformly illuminated slit can be approximated by a 1D function. In this case a single coordinate is sufficient to describe the illumination. However, in general, images in optics are bidimensional, so, the Fourier transform concepts introduced in the previous chapter must be extended to 2D.

The Fourier transform equation (equation (3.10)) can be easily extended to 2D as:

$$F(\xi, \eta) = \int_{-\infty}^{\infty} \int_{-\infty}^{\infty} f(x, y) \exp(-i2\pi(\xi x + \eta y)) dx dy \qquad (4.1)$$

Here, $f(x, y)$ is the 2D function whose Fourier transform (see section 1.4) is to be calculated, and ξ and η are the spatial frequencies along the two axes of the

doi:10.1088/978-0-7503-6392-1ch4

frequency domain. The inverse Fourier transform (equation (3.9)) will transform to:

$$f(x, y) = \int_{-\infty}^{\infty} \int_{-\infty}^{\infty} F(\xi, \eta) \exp(i2\pi(\xi x + \eta y)) \mathrm{d}\xi \mathrm{d}\eta \qquad (4.2)$$

In this case, the function $f(x, y)$ is decomposed into 2D harmonics of the form $\exp(i2\pi(\xi x + \eta y))$ and $F(\xi, \eta)\mathrm{d}\xi\mathrm{d}\eta$ represents the amplitude of the harmonic with spatial frequencies ξ and η.

4.1.1 The meaning of exponential harmonic functions in 2D. Spatial frequencies

To illustrate the meaning of the 2D harmonic consider the harmonic:

$$f(x, y) = \sin(2\pi x) \qquad (4.3)$$

This function[1] has a spatial frequency of 1. Note that while it is defined as a 2D function, it does not depend on y: for a given value of x, the function has the same value regardless of the value of y. A 3D surface plot of this function in the range 0–5 for both x and y, is shown in the left panel of figure 4.1 and the corresponding contour plot in the right panel.

At the top of the contour plot, the dependence of the function on x for a given value of y (in this case 2.5), is shown. As expected from the equation of the harmonic ($\xi = 1$), the spatial period along this axis is equal to 1. On the right side of the contour plot, the dependence of a function with y for $x = 2.5$ is shown. The spatial frequency η is zero, meaning the function is constant for a given value of x. The contour lines for this function are parallel to the x-axis. For harmonic functions with constant amplitude, contour lines will coincide with equal-phase lines, since the value of the function depends on the phase (up to a real or complex constant). In what follows we will use terms interchangeably.

Let us now calculate the Fourier transform for the general function $\sin(2\pi\xi_0 x)$ using equation (4.1).

$$F(\xi, \eta) = \int_{-\infty}^{\infty} \int_{-\infty}^{\infty} \sin(2\pi\xi_0 x) \exp(-i2\pi(\xi x + \eta y)) \mathrm{d}x\mathrm{d}y$$

Using the Euler relation:

$$F(\xi, \eta) = \int_{-\infty}^{\infty} \int_{-\infty}^{\infty} \frac{1}{2i}(\exp(i2\pi(\xi_0 x)) - \exp(-i2\pi(\xi_0 x)))\exp(-i2\pi(\xi x + \eta y))\mathrm{d}x\mathrm{d}y$$

$$= \frac{1}{2i} \int_{-\infty}^{\infty} \int_{-\infty}^{\infty} [\exp(i2\pi(\xi_0 x))\exp(-i2\pi(\xi x + \eta y))$$

$$- \exp(-i2\pi(\xi_0 x)) \exp(-i2\pi(\xi x + \eta y))]\mathrm{d}x\mathrm{d}y$$

---

[1] In the complex exponential notation this harmonic should be written as: $Im(\exp(i2\pi x))$ because $\exp(i2\pi x) = \cos(2\pi x) + i\sin(2\pi x)$ according to the Euler formula.

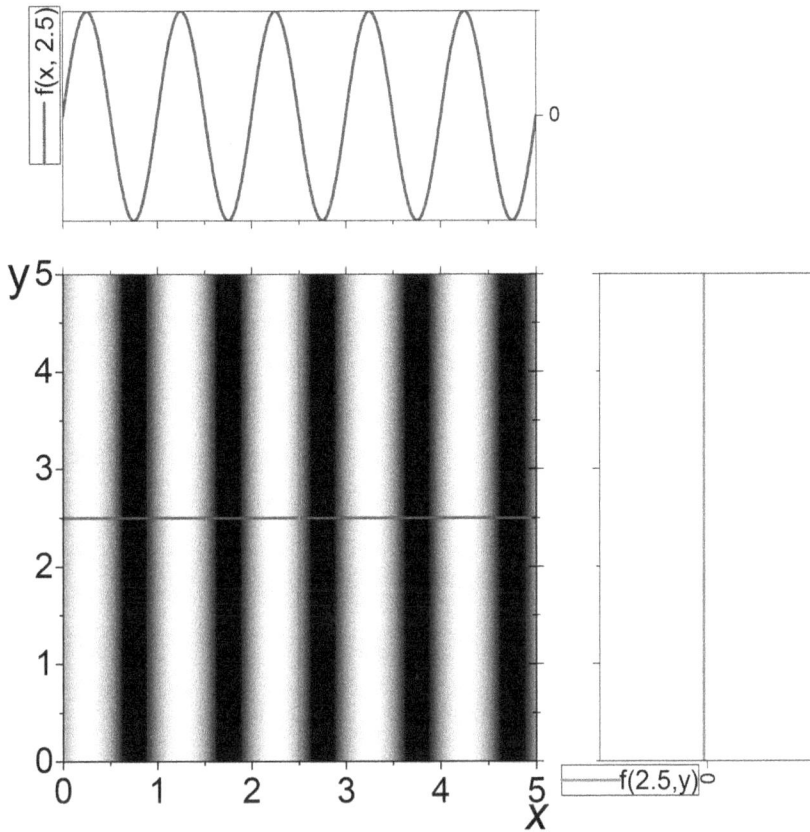

Figure 4.1. In the left panel a plot of the harmonic function $f(x, y) = \sin(2\pi x)$. In the right panel, the contour plot of the function is presented. The form of $f(x, 2.5)$ and $f(2.5, y)$ are shown alongside the contour plot. As expected, the spatial frequency in x ξ is 1 while the spatial frequency η in y is zero.

Simplifying the integrand:

$$= \frac{1}{2i} \int_{-\infty}^{\infty} \int_{-\infty}^{\infty} [\exp(i2\pi(\xi_0 - \xi)x) \exp(-i2\pi\eta y)$$
$$- \exp(-i2\pi(\xi_0 + \xi)x) \exp(-i2\pi\eta y)]\mathrm{d}x\mathrm{d}y$$
$$= \frac{1}{2i} \int_{-\infty}^{\infty} \exp(i2\pi(\xi_0 - \xi)x)\mathrm{d}x \int_{-\infty}^{\infty} \exp(-i2\pi\eta y)\mathrm{d}y$$
$$- \frac{1}{2i} \int_{-\infty}^{\infty} \exp(-i2\pi(\xi_0 + \xi)x)\mathrm{d}x \int_{-\infty}^{\infty} \exp(-i2\pi\eta y)\mathrm{d}y$$

Using equation (3.21):

$$= \frac{1}{2i}[\delta(\xi_0 - \xi)\delta(\eta)] - \frac{1}{2i}[\delta(\xi_0 + \xi)\delta(\eta)] \tag{4.4}$$

Considering, as in chapter 3, that $\exp(i\pi/2) = i$ and $\exp(i\pi) = -1$, and using the fact that delta functions are even, equation (4.4) becomes:

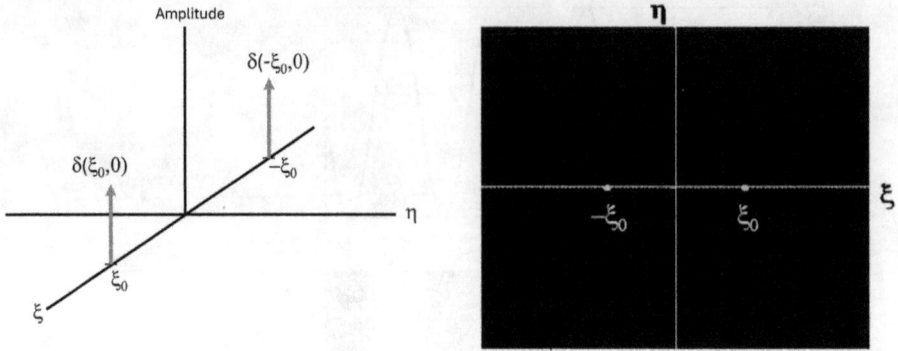

Figure 4.2. Graphic representation of the Fourier transform of equation (4.5) in a 3D plot (left panel) and in a grey scale contour map (right panel).

$$F(\xi, \eta) = \frac{1}{2}[\exp(-i\pi/2)\delta(\xi - \xi_0)\delta(\eta)] + [\exp(i\pi/2)\delta(\xi + \xi_0)\delta(\eta)] \qquad (4.5)$$

Then, the amplitudes of the two terms are $\frac{1}{2}\delta(\xi - \xi_0)\delta(\eta)$ and $\frac{1}{2}\delta(\xi + \xi_0)\delta(\eta)$ and their phases are $-\pi/2$ and $\pi/2$, respectively.

The products of the delta functions in equation (4.5) represent 2D delta functions. For example, the product $\delta(\xi - \xi_0)\delta(\eta)$, is nonzero only at $\xi = \xi_0$ and $\eta = 0$, meaning it is a 2D delta function located at the point $(\xi_0, 0)$. Similarly, $\delta(\xi + \xi_0)\delta(\eta)$ is a 2D delta function located at $(-\xi_0, 0)$.

Figure 4.2 shows the amplitude spectrum of equation (4.5) in a 3D plot (left panel) and in a contour map (right panel). In this contour map, the maximum value of the function is $\frac{1}{2}$ (represented as white) and the minimum value is zero (represented as black). In this case only two different shades are needed, and a one-bit pixel depth is sufficient for this black-and-white image. Many software packages can calculate and represent Fourier transforms in 3D surface plots or contour maps. It is important to consider that these contour maps often represent the square of the Fourier transform (referred to as the power spectrum) and are represented in logarithmic scales.

Analytically, sinusoidal functions extend from $-\infty$ to ∞, but real-world signals or images with sinusoidal shapes cannot extend infinitely. Thus, as shown in the solution of problem 3.9, the delta functions must be replaced by sinc functions centred at $(-\xi_0, 0)$ and $(\xi_0, 0)$. This will be reflected in the contour plot by the spreading of the dots.

In the more general case, the harmonic functions have both frequencies ξ and η different from zero and the equal-phase lines are oriented in a direction which is not perpendicular to either axis. The reader can easily demonstrate (using calculations like those that led to equation (4.5)) that the Fourier transform of $f(x, y) = \sin(2\pi\xi_0 x + 2\pi\eta_0 y)$ is:

$$F(\xi, \eta) = \frac{1}{2}\Big[\exp(-i\pi/2)\delta(\xi - \xi_0)\delta(\eta - \eta_0)\Big] + \frac{1}{2}\Big[\exp(i\pi/2)\delta(\xi + \xi_0)\delta(\eta + \eta_0)\Big]$$

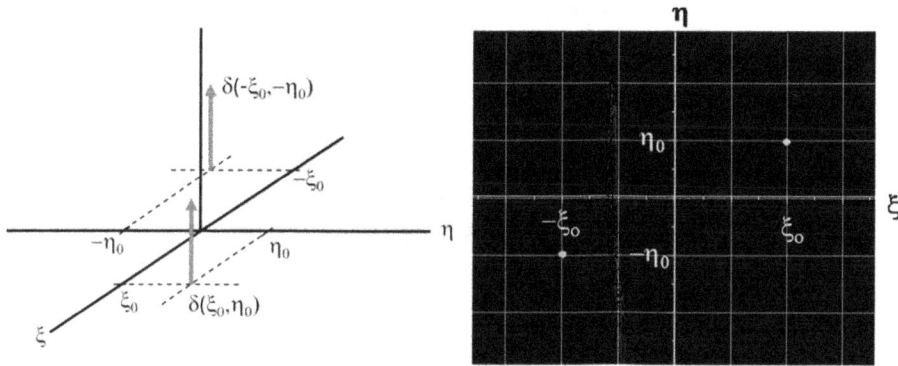

Figure 4.3. The amplitude spectrum of the function $f(x, y) = \sin(2\pi\xi_0 x + 2\pi\eta_0 y)$ using a 3D surface plot (left) and a contour map (right).

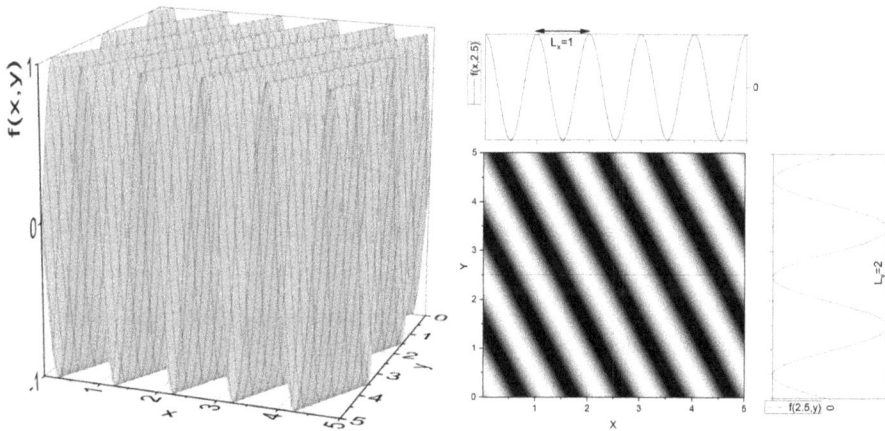

Figure 4.4. In the left panel a plot of the harmonic function $f(x, y) = \sin(2\pi x + \pi y)$. In the right panel, the contour map of the function is presented. The form of $f(x, 2.5)$ and $f(2.5, y)$ are shown alongside the contour plot. As expected, the spatial frequency ξ in x is 1, while the spatial frequency η in y is 0.5.

Graphically, the amplitudes for the two frequencies are represented in figure 4.3 using a 3D surface plot (left) and a contour map (right).

An example of such a harmonic is shown in figure 4.4. In this case, for the function:

$$f(x, y) = \sin(2\pi x + \pi y) \tag{4.6}$$

with $\xi_0 = 1$ and $\eta_0 = 0.5$, a 3D surface plot for this function is presented in the left panel of figure 4.4. In the right panel, the corresponding contour map shows that the equal-phase lines are neither parallel to the x-axis nor to the y-axis. The dependence on x for a given y value and the dependence on y for a given x value are also shown, with the spatial periods L_x and L_y indicated. It can be noted that $L_x = 1/\xi_0 = 1$ and $L_y = 1/\eta_0 = 2$ as expected.

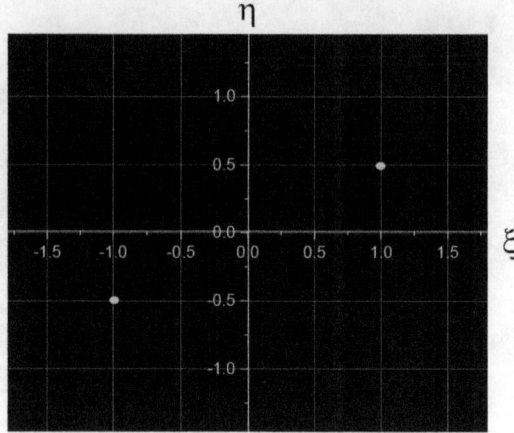

Figure 4.5. Contour map of the amplitude of the Fourier transform of the function $f(x, y) = \sin(2\pi x + \pi y)$.

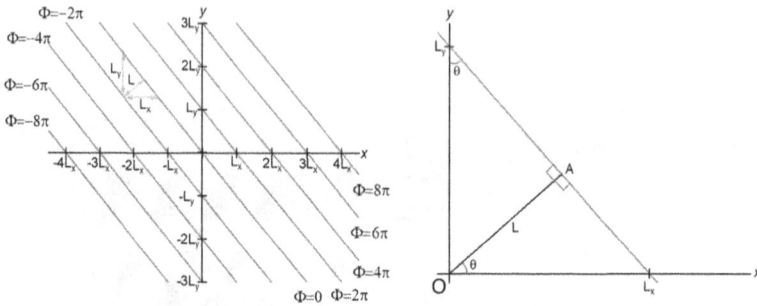

Figure 4.6. Equal-phase lines for a 2D harmonic with spatial periods L_x and L_y. On the right, only the $\phi = 2\pi$ curve is displayed in the first quadrant.

In this case, the contour plot will present dots (assuming an infinite extent of the harmonic) at (ξ_0, η_0) and $(-\xi_0, -\eta_0)$, that is $(1, 0.5)$ and $(-1, -0.5)$ in the (ξ, η) plane of the frequency domain. This spectrum is presented in figure 4.5.

It is worth noting that the delta functions in the contour map are aligned perpendicularly to the contour lines (assuming that the axes x, y in the spatial domain, are aligned with the axes ξ, η in the frequency domain, respectively). As it will be explained below, this fact is useful for quickly identifying, which points in the frequency domain correspond to periodic features in the space domain.

4.1.2 Equal-phase lines for 2D harmonic functions

In the left panel of figure 4.6, some equal-phase lines (parallel to contour lines as defined earlier) are represented for phases that are multiples of 2π for a general 2D harmonic function $\exp(i2\pi(\xi_0 x + \eta_0 y))$. The spatial periods L_x and L_y, as well as the spatial period of the function as a whole, L, are indicated. Let's call ϕ the phase and we have:

$$\phi = 2\pi(\xi_0 x + \eta_0 y)$$

or:

$$\phi = 2\pi\left(\frac{1}{L_x}x + \frac{1}{L_y}y\right)$$

For $\phi = 0$:

$$2\pi\left(\frac{1}{L_x}x + \frac{1}{L_y}y\right) = 0$$

$$\left(\frac{1}{L_x}x + \frac{1}{L_y}y\right) = 0$$

And the equation for the corresponding line is:

$$y = -\frac{L_y}{L_x}x$$

It is the equation for a straight line with negative slope and passing through the origin as represented in the left panel of figure 4.6. For $\phi = 2\pi$ it is easy to obtain the equation:

$$y = L_y - \frac{L_y}{L_x}x$$

Which intersects the x- and y-axes in L_x and L_y, respectively. The other lines in figure 4.6 can be reproduced using the ϕ values indicated in the figure.

In the right panel of figure 4.6 the equal-phase line corresponding with $\phi = 2\pi$ is considered. The angles θ represented in the figure are equal because their sides are perpendicular. This is due to the definition of point A, such that OA is the height of the triangle $\Delta L_y OL_x$. For the triangle ΔAOL_x, we can write:

$$\cos(\theta) = \frac{L}{L_x} \tag{4.7}$$

And for the triangle $\Delta L_y OL_x$:

$$\cos(\theta) = \frac{L_y}{\sqrt{L_x^2 + L_y^2}} \tag{4.8}$$

And equating (4.7) and (4.8) follows:

$$L = \frac{1}{\sqrt{\frac{1}{L_x^2} + \frac{1}{L_y^2}}} = \frac{1}{\sqrt{\xi_0^2 + \eta_0^2}} \tag{4.9}$$

That means that a harmonic function with spatial periods L_x and L_y (spatial frequencies ξ_0 and η_0) can be interpreted as a harmonic function with equal-phase lines obliques with respect to the x or y directions. Naturally, if one of the coordinate axes is chosen to be normal to these equal-phase lines, the function will have only one spatial frequency and one spatial period L.

When a 2D function is decomposed in 2D harmonics using the Fourier transform, these harmonics typically exhibit a continuous spectrum of frequencies ξ and η, and equal-phase lines oriented in various directions. As discussed earlier, the delta functions or sinc functions corresponding to each harmonic will be positioned at different points in the (ξ, η) plane. The contour map of the Fourier transform spectrum is, in general, a 2D image in the frequency domain.

4.2 Separable variables functions

A separable variable 2D function $f(x, y)$ is one that can be expressed as the product of two functions, each depending on only one variable:

$$f(x, y) = g(x)h(y) \tag{4.10}$$

Whether a function is separable depends on the function itself but also on the chosen coordinate system.

For separable functions, we can use equations (4.1) and (4.10) and write:

$$F(\xi, \eta) = \int_{-\infty}^{\infty} \int_{-\infty}^{\infty} f(x, y) \exp(-i2\pi(\xi x + \eta y)) \mathrm{d}x \mathrm{d}y$$

$$= \int_{-\infty}^{\infty} \int_{-\infty}^{\infty} g(x)h(y) \exp(-i2\pi(\xi x + \eta y)) \mathrm{d}x \mathrm{d}y$$

This simplifies to:

$$F(\xi, \eta) = \int_{-\infty}^{\infty} g(x) \exp(-i2\pi\xi x) \mathrm{d}x \int_{-\infty}^{\infty} h(y) \exp(-i2\pi\eta y) \mathrm{d}y$$
$$= F\{g(x)\}F\{h(y)\} \tag{4.11}$$

In other words, the Fourier transform of the product of two functions, each depending on only one variable, is equal to the product of the Fourier transforms of the individual functions.

4.2.1 The 2D squared function

An example of a 2D separable function is the 2D rectangular pulse of side a in which:

$$h(x, y) = f(x, y)g(x, y) \tag{4.12}$$

Following the definitions for the 1D rectangular functions (equation (3.11)):

$$f(x, y) = \mathrm{Rect}_a(x) \text{ and } g(x, y) = \mathrm{Rect}_a(y)$$

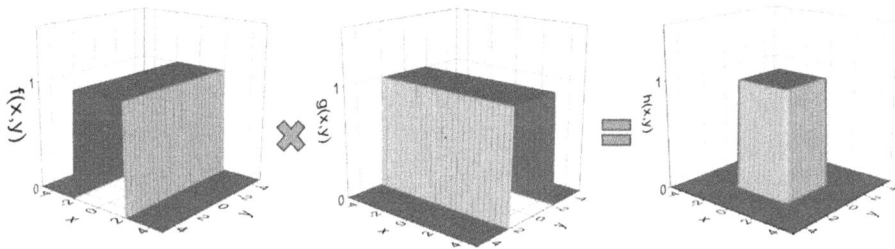

Figure 4.7. 2D square pulse as a product of the 1D square pulses in x and y.

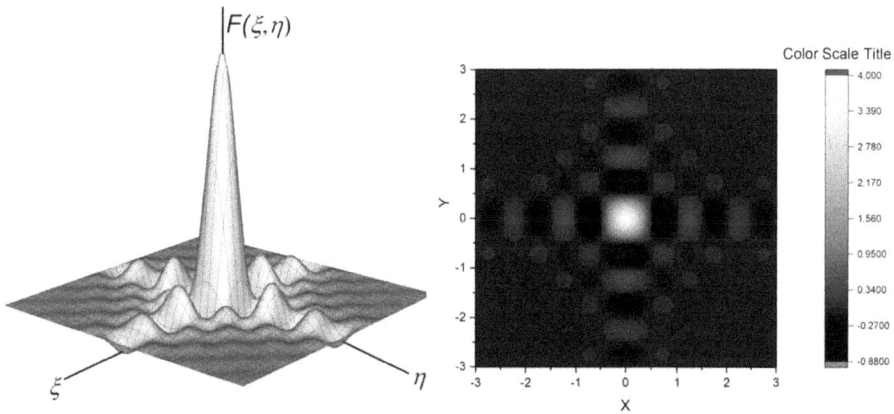

Figure 4.8. The Fourier transform of the 2D square pulse: surface plot (left), contour map (right).

In optics, this function can represent the illumination of a square aperture of side length a. It is illustrated in figure 4.7 as a product of the functions $\text{Rect}_a(x)$ and $\text{Rect}_a(y)$.

Using equation (3.12) (the Fourier transform of the Rect function), the Fourier transform of the 2D rectangular pulse is:

$$F(\xi, \eta) = a^2 \, \text{sinc}(\pi\xi a) \, \text{sinc}(\pi\eta a) \tag{4.13}$$

This Fourier transform is a 2D function in the frequency domain and is represented in a 3D surface plot and in a contour plot in figure 4.8.

4.3 Bidimensional images. Filtering. Use of 'Image J' program for image filtering

4.3.1 Bidimensional images

Consider again the image in the right panel of figure 4.1. It is a very simple image composed by only one harmonic. The Fourier transform reveals that any image, no matter how complex, can be represented as the sum (or, more generally, the integral) of 2D harmonics with appropriate frequencies and amplitudes.

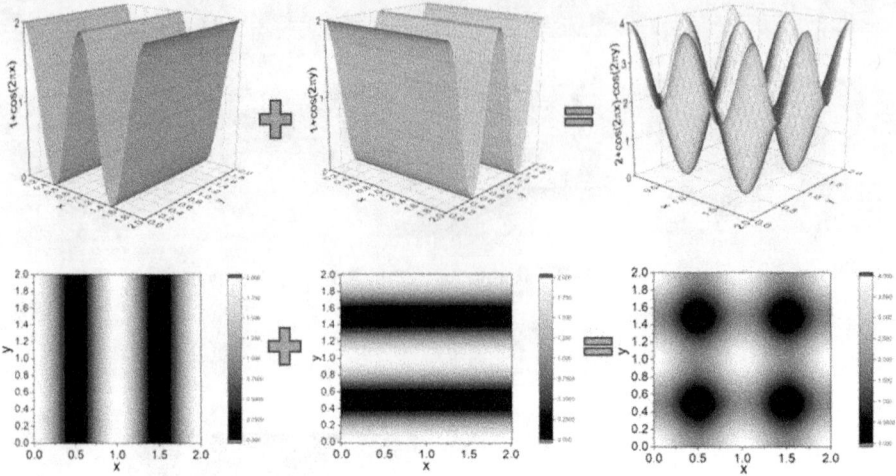

Figure 4.9. 3D surface (top) and contour (bottom) plots of the function represented in equation (4.14).

For example, consider the function:

$$f(x, y) = 2 + \cos(2\pi\xi_0 x) + \cos(2\pi\eta_0 y) \tag{4.14}$$

with $\xi_0 = \eta_0 = 1$. It is still a simple function but slightly more complex than, for example, equation (4.3) or even equation (4.6). It consists of a constant term and two harmonics with the same frequency but with equal-phase lines oriented in perpendicular directions[2]. The sum is represented in the upper panel of figure 4.9 in a 3D surface plot, and in the lower panel using a contour map.

It is interesting to show the frequency spectrum of this image. The Fourier transform of the function $f(x, y) = 2 + \cos(2\pi\xi_0 x) + \cos(2\pi\eta_0 y)$ is:

$$F\{f(x, y)\} = F\{2\} + F\{\cos(2\pi\xi_0 x)\} + F\{\cos(2\pi\eta_0 y)\}$$

And, according to problem 3.8 or using equation (3.25):

$$\begin{aligned} F\{f(x, y)\} = 2\delta(\xi)\delta(\eta) &+ \frac{1}{2}[\delta(\xi - \xi_0) + \delta(\xi + \xi_0)]\delta(\eta) \\ &+ \frac{1}{2}[\delta(\eta - \eta_0) + \delta(\eta + \eta_0)]\delta(\xi) \end{aligned} \tag{4.15}$$

That is, a delta function in the origin with area equal to 2; and delta functions centred at $(\xi_0, 0)$, $(-\xi_0, 0)$, $(0, \eta_0)$, and $(0, -\eta_0)$ in the (ξ, η) plane. This spectrum is shown in figure 4.10.[3]

---

[2] Since the function will represent a distribution of intensity, the constant (number 2) has been included to avoid negative values. By adding this number, the harmonic is vertically translated, and the minimum of the function corresponds to zero.

[3] In real images, harmonics do not extend in the whole range of the space domain, and their Fourier transform is not composed of delta functions but rather of sinc functions. This means that the white points in figure 4.10 will appear as spread-out regions in the frequency domain.

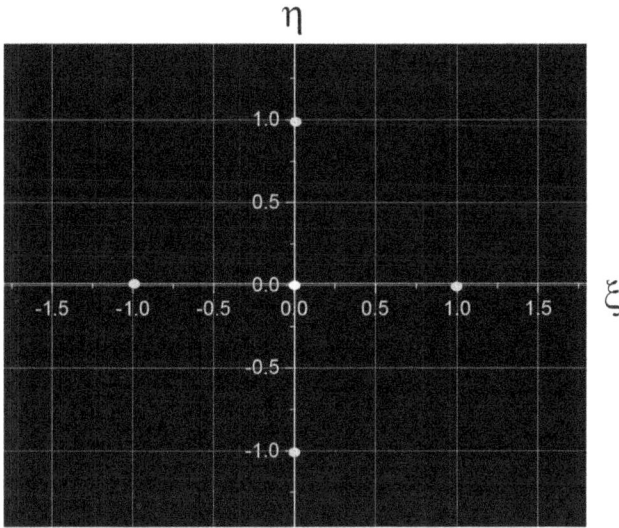

Figure 4.10. The Fourier spectrum of the function represented by equation (4.14) and figure 4.9. Note that the delta function at the origin has four times the area of the other. It is represented by a brighter point.

Figure 4.11. Left: A monochromatic version of the painting 'The girl with the pearl earring'; right: the square of the Fourier transform of the picture displayed in a logarithmic scale according to the Image J software. Girl with the Pearl Earring Original Image Credit: Pixabay by robertwaghorn.

In the case of more complex images, a much larger number of harmonics are needed. These harmonics will have equal-phase line oriented in different directions and varying frequencies. In general, the Fourier transform becomes a relatively complex function, and instead of isolated points in the frequency domain, intricate patterns are formed.

As an example of a complex image, figure 4.11 shows a monochromatic version of the famous painting 'The girl with the pearl earring' by Johannes Vermeer. Using the publicly available software **Image J**, the Fourier transform of the image was calculated. The right hand of figure 4.11 displays the square of this Fourier transform in a logarithmic scale. It can be observed that individual points in the spectrum are difficult to distinguish due to the vast number of harmonic functions needed to reproduce the image. The contribution of each harmonic overlap that of the other in different frequencies and directions creating a complex frequency spectrum.

The Image J program allows one to perform the inverse Fourier transform and recover the image from its Fourier transform. Readers can verify this by applying the process to any digital image. The operations in Image J are straightforward, and comprehensive tutorials are available on the web.

4.3.2 Filtering. Use of 'Image J' program for image filtering

An important feature of the Fourier transform is the ability to modify an image by altering its Fourier transform rather than the image itself. This process is called 'filtering'. Filtering involves removing (or filtering out) specific frequency regions in the frequency (or power) spectrum and then performing the inverse transform to obtain the modified image. The most common filtering processes are low-frequency filtering and high-frequency filtering.

4.3.2.1 Low-frequency filtering
In figure 4.12, the low-frequency filtering is applied to the image from figure 4.11 using Image J. The left panel shows the same spectrum as in figure 4.11, except that a small region near the origin of the frequency domain coordinate system (low frequencies) has been removed (represented by a black disk). The right panel shows

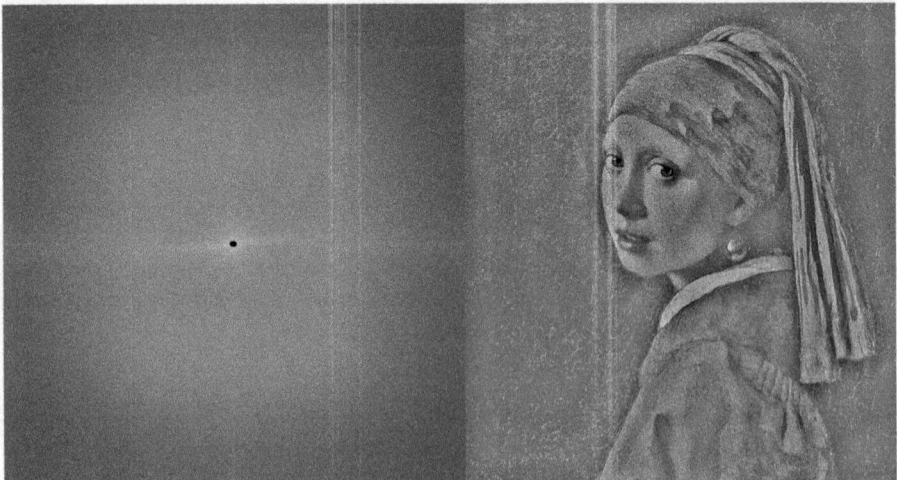

Figure 4.12. The process of low-frequency filtering. The low-frequency region in the power spectrum has been eliminated. The inverse Fourier transform provides an image with the contours emphasized. Girl with the Pearl Earring Original Image Credit: Pixabay by robertwaghorn.

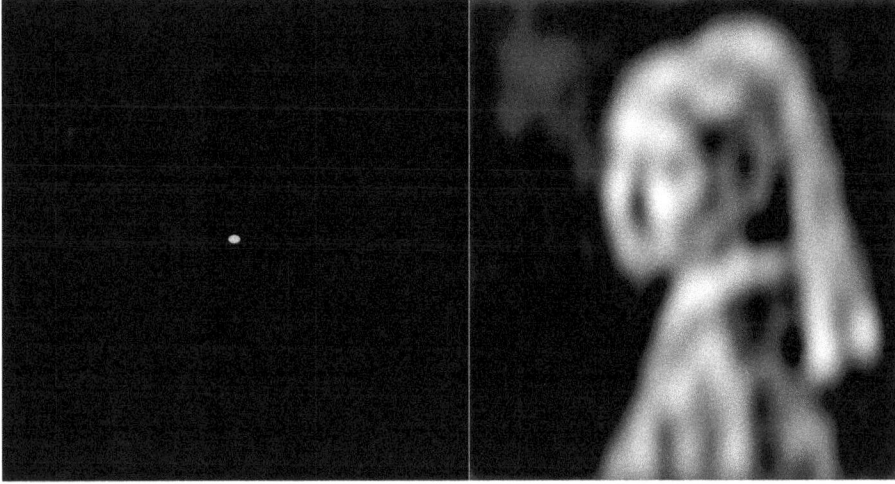

Figure 4.13. The process of high-frequency filtering. All frequencies have been eliminated except the low-frequency region in the power spectrum (bright disk). The inverse Fourier transform produces an image with blurred contour. Girl with the Pearl Earring Original Image Credit: Pixabay by robertwaghorn.

the result of applying the inverse Fourier transform to this modified spectrum. This is an example of low-frequency filtering. The effect is that the contours of shapes are emphasized, while regions with relatively constant illumination are darkened. This behaviour is expected because high frequencies are responsible for reproducing regions with abrupt changes in the image, as discussed in section 3.1.2. This technique can be used to enhance the contrast of an image.

4.3.2.2 High-frequency filtering
In figure 4.13, a high-frequency filter is demonstrated. Here, all the frequencies have been eliminated except the low frequencies near the origin (represented by a bright disk in the left panel). As shown on the right, after applying the inverse Fourier transform, the resulting image has blurred contours due to the absence of high frequencies in the power spectrum.

4.4 Fourier transform in polar coordinates (Fourier–Bessel transform)

In optics, functions with circular symmetry are of great importance. Lenses and circular apertures are often encountered in optical systems, and to analyse such objects, it is convenient to express the Fourier transform in polar coordinates. Starting from the expression:

$$F(\xi, \eta) = \int_{-\infty}^{\infty} \int_{-\infty}^{\infty} f(x, y) \exp(-i2\pi(\xi x + \eta y)) \mathrm{d}x \mathrm{d}y$$

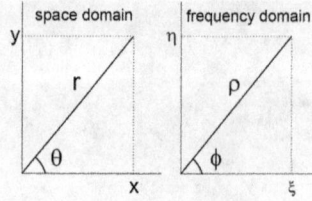

Figure 4.14. Transformation of coordinates in both the space and frequency domains.

we must transform both the spatial and frequency coordinates. Figure 4.14 illustrates this coordinate transformation. (x, y) must be transformed in (r, θ) and (dx, dy) in $(dr, d\theta)$. The transformation equations are:

$$x = r\cos\theta; \; y = r\sin\theta; \; dxdy = rdrd\theta \tag{4.16}$$

for the space domain, and

$$\xi = \rho\cos\phi; \; \eta = \rho\sin\phi; \; d\xi d\eta = \rho d\rho d\phi \tag{4.17}$$

for the frequency domain.

Substituting equations (4.16) and (4.17) in (4.1) we obtain:

$$F(\rho, \phi) = \int_0^\infty \int_0^{2\pi} f(r, \theta)\exp(-i2\pi(\rho r\cos\theta\cos\phi + \rho r\sin\theta\sin\phi))rdrd\theta \tag{4.18}$$

In many cases, the function $f(r, \theta)$ has spheric symmetry (i.e. does not depend on θ), so we can write $f(r, \theta) \equiv f(r)$. Consequently, $F(\rho, \phi) \equiv F(\rho)$. Using the trigonometric identity: $\cos\theta\cos\phi + \sin\theta\sin\phi = \cos(\theta - \phi)$ (see section 1.2.3), we have:

$$F(\rho, \phi) = \int_0^\infty \int_0^{2\pi} rf(r)\exp(-i2\pi\rho r\cos(\theta - \phi))drd\theta$$

Since $F(\rho, \phi) \equiv F(\rho)$ the integral will not depend on ϕ and we can simplify by setting $\phi = 0$. Thus:

$$F(\rho) = \int_0^\infty rf(r)\left[\int_0^{2\pi} \exp(-i2\pi\rho r\cos(\theta))d\theta\right]dr \tag{4.19}$$

The definition of the zero-order Bessel function (see equation (1.17)), that is an even function, as we can see in the left panel of figure 1.12(a):

$$J_0(-u) = J_0(u) = \frac{1}{2\pi}\int_0^{2\pi} \exp(-iu\cos(\theta))d\theta \tag{4.20}$$

Letting $u = 2\pi\rho r$

$$J_0(2\pi\rho r) = \frac{1}{2\pi}\int_0^{2\pi} \exp(-i2\pi\rho r\cos(\theta))d\theta$$

Substituting this into equation (4.18):

$$F(\rho) = \int_0^\infty rf(r)2\pi J_0(2\pi\rho r)dr \qquad (4.21)$$

Let us apply this result to the case of a circular aperture. Consider the cylindric function defined as:

$$f(r) = \begin{cases} 0 \text{ if } r > a \\ 1 \text{ if } r \geqslant a \end{cases}$$

In optics, this function can represent a circular hole uniformly illuminated. For this function, equation (4.21) becomes:

$$F(\rho) = \int_0^a r2\pi J_0(2\pi\rho r)dr \qquad (4.22)$$

We can now use equation (1.16):

$$aJ_1(a) = \int_0^a uJ_0(u)du \qquad (4.23)$$

We can rewrite equation (4.22) in a form compatible with equation (4.23). Multiplying equation (4.22) by $(\frac{2\pi\rho}{2\pi\rho})^2$, we obtain:

$$F(\rho) = \frac{2\pi}{(2\pi\rho)^2} \int_0^a 2\pi\rho r J_0(2\pi\rho r)d(2\pi\rho r)$$

Letting $u = 2\pi\rho r$, this becomes:

$$F(\rho) = \frac{2\pi}{(2\pi\rho)^2} \int_0^{2\pi\rho a} uJ_0(u)du$$

Using equation (4.23), we find:

$$F(\rho) = \frac{2\pi}{(2\pi\rho)^2} \; 2\pi\rho a J_1(2\pi\rho a) = \frac{a}{\rho}J_1(2\pi\rho a) = 2\pi a^2\frac{J_1(2\pi\rho a)}{2\pi\rho a}$$

Thus:

$$F(\rho) = 2A\frac{J_1(2\pi\rho a)}{2\pi\rho a} \qquad (4.24)$$

where $A \equiv \pi a^2$ is the area of the circular hole.

In figure 4.15, the zero (top) and first (middle) order Bessel functions ($J_0(u)$ and $J_1(u)$) as well as $\frac{J_1(u)}{u}$ (bottom) are plotted as a function of u. The values of the first five zeros of the functions are indicated.

The image of a circular aperture, as well as the square of its Fourier transform, calculated using the Image J software are shown in figure 4.16.

Equation (4.24) is known as the Airy pattern in honour of George Biddell Airy, who mathematically described the diffraction of light through a circular aperture. This pattern is fundamental in optics and astronomy, as it describes how images are

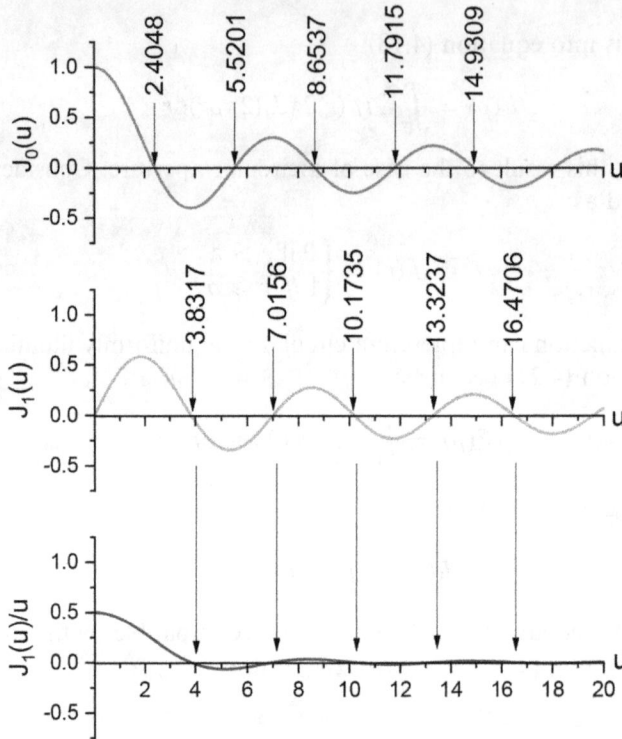

Figure 4.15. Zero (top) and first (middle) order Bessel functions ($J_0(u)$ and $J_1(u)$), as well as $\frac{J_1(u)}{u}$ (bottom). The values of the first five zeros of the functions are indicated.

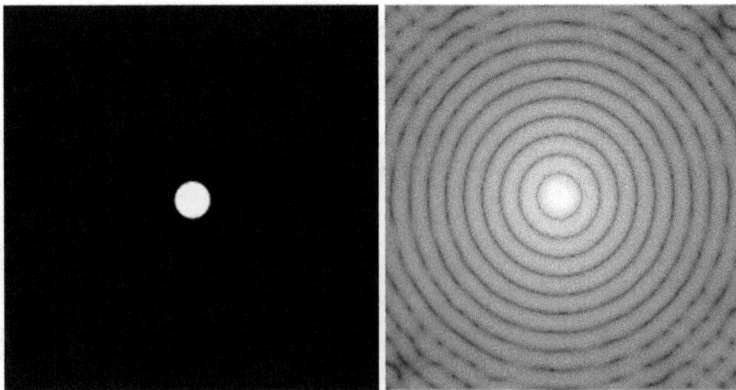

Figure 4.16. A circular aperture and the square of its Fourier transform obtained from the Image J software.

formed in optical systems and sets resolution limits for instruments such as telescopes and microscopes. The first zero of the pattern occurs when $2\pi\rho a = 3.8317$. Then:

$$\rho = \frac{3.8317}{2\pi a} = \frac{1.22}{2a} \tag{4.25}$$

The Airy pattern is also related to the Rayleigh criterion, which determines when two light sources (such as two stars) can be resolved as separate. According to this criterion: two points of light are 'just resolved' when the central maximum of the Airy pattern of one coincides with the first minimum of the Airy pattern of the other.

To deduce this criterion from the above relation we have to consider that the radial spatial frequency ρ is $\frac{1}{\lambda} \sin \alpha$, where α is the angular distance in the plane of observation. For small angles:

$$\rho = \frac{1}{\lambda}\alpha \tag{4.26}$$

According to the Rayleigh criterion, the distance between the objects must correspond with the radial spatial frequency in which $F(\rho) = 0$, which was calculated as equation (4.25). Equating (4.25) and (4.26):

$$\alpha = 1.22\frac{\lambda}{2a}$$

which is the Rayleigh criterion.

4.5 Problems

Problem 4.1 Consider the harmonic 2D function $\exp{(i(0.5\pi x + 0.2\pi y))}$ (with x and y in cm).
   (a) What is the direction of the normal to the equal-phase lines? (give an angle)
   (b) What is the spatial period and the spatial frequency of the function as a whole?
   (c) Plot the equal-phase-lines in the x,y plane.
   (d) Calculate the Fourier transform of this function and represent it graphically in a ξ, η plane.

Problem 4.2 Represent the frequency spectrum of the function:

$$f(x, y) = 1 + \cos{(2\pi x + 6\pi y)} + 1 + \cos{(\pi x + 4\pi y)} + 1 + \cos{(3\pi x)}$$

Further Reading

[1] Goodman J W 2017 *Introduction to Fourier Optics* 4th edn (New York: W.H. Freeman)
[2] Peters T M and Parker K J 1998 *The Fourier Transform in Biomedical Engineering* (Boston, MA: Birkhäuser)
[3] Bracewell R N 1995 *Two-Dimensional Imaging* (Upper Saddle River, NJ: Prentice-Hall)
[4] Bracewell R N 2003 *Fourier Analysis and Imaging* (New York: Springer)
[5] Petrou M and Petrou C 2010 *Image Processing: The Fundamentals* 2nd edn (Hoboken, NJ: Wiley)

IOP Publishing

Elementary Fourier Optics for Science and Engineering Students

Hélène Ollivier and Osvaldo de Melo

Chapter 5

Linear and shift-invariant systems

This chapter explores linear and space-invariant systems, focussing on how images can be interpreted as the convolution of an object with the point spread function (PSF) of an optical system. It discusses fundamental concepts—including system linearity, spatial invariance, point sources, and the PSF—and highlights their significance for image processing and analysis. A graphical method for performing convolution in both 1D and 2D systems is explained. The chapter also covers essential properties of convolution, such as its operation with the delta function, commutativity, and the convolution theorem. To reinforce these concepts, the chapter concludes with proposed problems on convolution and linear systems.

5.1 Linear systems. Input and outputs. Examples

A system is a set of interconnected components that collectively transform an input function into an output function. This transformation may be governed by a mathematical equation, an algorithmic process, or other operational rules. For instance, a system could simply multiply the input function by a constant factor. When this constant exceeds unity, the system acts as an *amplifier*. The constant may also be complex, in which case it introduces a phase shift to the input signal.

In optical systems, the input typically represents the electric field distribution (or light intensity) at the object plane, while the output corresponds to the transformed distribution at the image plane. Such systems may consist of individual components (e.g., lenses, mirrors, prisms, polarizers) or combinations thereof, each contributing to the overall transformation.

The action of a system S can be mathematically represented as:

$$S\{f(x)\} = g(x) \tag{5.1}$$

where $f(x)$ denotes the input function and $g(x)$ the output function. Consider two specific input–output pairs:

$$S\{f_1(x)\} = g_1(x) \text{ and } S\{f_2(x)\} = g_2(x)$$

The system S is classified as *linear* if it satisfies the following condition for any two inputs $f_1(x)$ and $f_2(x)$:

$$S\{a_1 f_1(x) + a_2 f_2(x)\} = a_1 g_1(x) + a_2 g_2(x) \tag{5.2}$$

where a_1 and a_2 are (in general complex) coefficients.

In simple terms: 'the output of a linear combination of inputs is simply the same linear combination of individual outputs.' This is equivalent to the principle of superposition, and it is an approximation that simplifies the solution of many problems. Linearity means that the action of the system does not depend on the intensity of the input, or that the action over an input does not depend on the existence of other inputs. It can be noted that linearity implies that: (i) the constants of the linear combination can be extracted from the braces; and (ii) the outputs add up.

5.2 Shift-invariant systems

If $S\{f(x)\} = g(x)$, then the system is said to be shift-invariant if:

$$S\{f(x - x_0)\} = g(x - x_0).$$

For any input shift x_0. This means that: when the input function is displaced by x_0, the output undergoes an identical displacement while its shape remains unchanged.

5.3 Decomposition in delta functions

In optics, when the input (object) is approximated by a point source—mathematically represented by a 2D delta function in the object plane—an ideal optical system would produce an equally perfect point-like image. Thus, an object composed of multiple points would yield an exact collection of corresponding image points, forming a mathematically perfect reproduction. However, real optical systems cannot achieve this ideal behaviour due to inherent aberrations (geometric imperfections in lenses/mirrors) and diffraction effects (caused by finite apertures). These limitations result in a blurred image of each point source, where the degree of spreading is characterized by the system PSF[1]. Consequently, even for an ideal point object, the actual image is never a perfect point but rather a localized intensity distribution, imposing fundamental limits on resolution and image fidelity.

In optics, we can model any object as a collection of point sources represented by delta functions. Let us recall that $\int_{-\infty}^{\infty} \delta(x - x_0)dx = 1$, and obviously $\int_{-\infty}^{\infty} 2\,\delta(x - x_0)dx = 2$. In this case the number 2 is the coefficient which scales the delta function area (see section 3.4.1). Also, using the sifting property of the delta function:

---

[1] From the use in time dependent functions in electronics, the point object is also called 'impulse' and the PSF, 'impulse response'.

$$\int_{-\infty}^{\infty} f(x)\, \delta(x - x_0)\mathrm{d}x = f(x_0) \tag{5.3}$$

To generalize this to variable positions (replacing x_0 with x), we introduce the dummy variable α (see section 3.4), yielding:

$$\int_{-\infty}^{\infty} f(\alpha)\, \delta(\alpha - x)\mathrm{d}\alpha = f(x) \tag{5.4}$$

Note that this equation can be interpreted as a linear combination of delta functions, where the coefficients correspond to the function's values at each point. In other words, using equation (5.4), we have decomposed $f(x)$ into an integral of delta functions weighted by the function's value at every point. At first glance, this decomposition of $f(x)$ into delta functions weighted by its own values (via Eq. 5.4) might appear almost trivial. However, its true power and subtlety become evident when applied to image formation, as we will explore in this section.

In optics, when the function represents the illumination distribution in the object plane, equation (5.4) shows that the image can be modelled as a collection of delta functions weighted by the illumination at each point. Figure 5.1 (right panel) illustrates this decomposition for a 1D line (marked by black points) from the image in the left panel. The figure displays representative weighted delta functions that collectively form the intensity distribution along the black line. This serves as an example of decomposing a 1D image into a linear combination of weighted deltas. Of course, for an exact reconstruction of the intensity distribution, these delta functions would need to be spaced infinitesimally close to each other.

In 2D, the decomposition is written as:

$$f(x, y) = \iint_{-\infty}^{\infty} f(\alpha, \beta)\delta(\alpha - x)\delta(\beta - y)\mathrm{d}\alpha\mathrm{d}\beta \tag{5.5}$$

Where the double integral runs over the area of the image. It can be noted that $\delta(\alpha - x)\delta(\beta - y)$ is a 2D delta function which represents a point source located in $(x, y) = (\alpha, \beta)$ in the object plane.

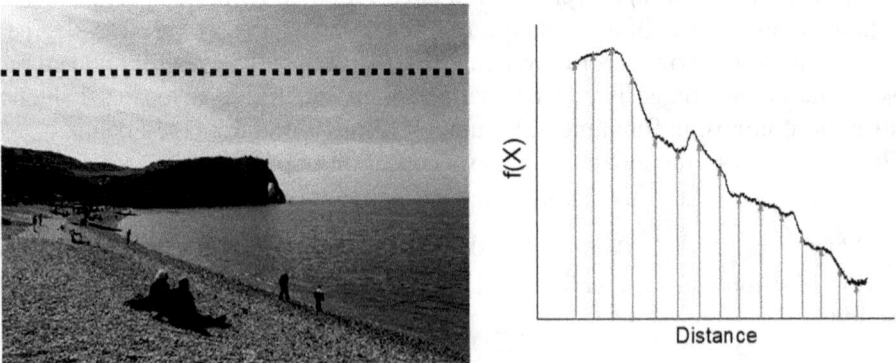

Figure 5.1. Illustration of the decomposition of a 1D function (dot line in the left panel) in a collection of delta functions weighted by the function values at every point (right panel).

5.4 Point spread function. The convolution integral. The image as a convolution

To analyse how an optical system modifies the illumination function of an object, let us consider a simplified 1D case: an illuminated line. The object's illumination is described by a (generally non-uniform) function $f(x)$ (e.g., the line in figure 5.1). Let $S: f(x) \rightarrow g(X)$ be a linear optical system that maps object plane functions to image plane functions, so that we can write:

$$g(X) = S\{f(x)\}$$

where $g(X)$ is the function describing the illumination of the image (note that X represents the coordinate in the image plane). Using equation (5.4), we decompose $f(x)$ into a superposition of weighted delta functions (as discussed above):

$$g(X) = S\left\{\int_{-\infty}^{\infty} f(\alpha)\,\delta(\alpha - x)\mathrm{d}\alpha\right\}$$

Here, $f(\alpha)$ is the weighting coefficient and, assuming linearity of the optical system:

$$g(X) = \int_{-\infty}^{\infty} f(\alpha)S\{\delta(\alpha - x)\}\mathrm{d}\alpha$$

$S\{\delta(\alpha - x)\}$ is the effect of the system on the point source located at $x = \alpha$. In 2D:

$$g(X, Y) = \iint_{-\infty}^{\infty} f(\alpha, \beta)S\{\delta(\alpha - x)\delta(\beta - y)\}\mathrm{d}\alpha\mathrm{d}\beta$$

Note that the system S turns a function that depends on the object point sources positions (x and y) into a function that depends on the image points positions (X and Y), which explains why the left hand side of the equation depends on X and Y while the right hand side is expressed as a function of x and y. The reader will realize that $S\{\delta(\alpha - x)\delta(\beta - y)\}$ is what we defined earlier as the PSF.

Let us suppose that the considered optical system is shift-invariant in space along the x and y directions. That means that the image has the same orientation and size as the object (it is not inverted nor magnified). Let us also call $h(X)$ the PSF of a point source $\delta(x)$ at the origin of the object plane: $h(X) = S\{\delta(x)\}$. Then, the PSF originating from a point source located in other x values, that is $\delta(\alpha - x)$, will be expressed as the function h centred in $X = \alpha$ in the image plane: $\text{PSF} \equiv S\{\delta(\alpha - x)\} = h(X - \alpha)$. In other words, each point source located in a given position α in the object plane, will give rise to a PSF shaped as h, centred at the same position α of the image plane. Then:

$$g(X) = \int_{-\infty}^{\infty} f(\alpha)h(X - \alpha)\mathrm{d}\alpha \tag{5.6}$$

In 2D:

$$g(X, Y) = \int_{-\infty}^{\infty} f(\alpha, \beta)h(X - \alpha, Y - \beta))\mathrm{d}\alpha\mathrm{d}\beta \tag{5.7}$$

Since the image of the point source is extended in X, the illumination of any point source will affect a relatively extended region in X, and the illumination of the different point sources will overlap in the image plane. Then, for calculating

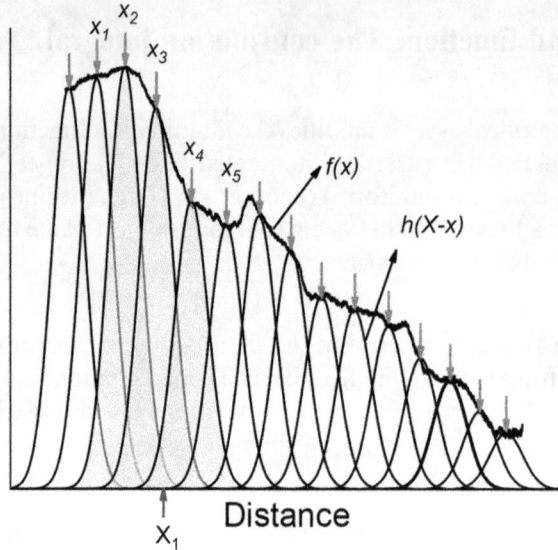

Figure 5.2. Illustration of the formation of the image from the contribution of the PSF of the source points.

the illumination in the image plane in a given point X_1, we will have to add not only the illumination coming form $x = X_1$ but also that coming from an extended region in x. This is the meaning of equation (5.6). This is illustrated in figure 5.2 in which a Gaussian type PSF was considered. To obtain the value of the illumination at the point X_1, for example, it is necessary to add the contributions of the point sources in the x range between x_1 and x_5: this is what equation (5.6) does for any value of X. Note that the PSFs are modulated by the function $f(x)$ and that the result of the integral in equation (5.6) is a function of X. The operation represented by this integral is called convolution. So $\int_{-\infty}^{\infty} f(x)h(X-x)\mathrm{d}x$ is the convolution of the function $f(x)$ and $h(x)$, symbolically: $(f \otimes h)(X)$.[2] It can be noted that the convolution depends on X since x is a dummy variable in the convolution integral.

So:

$$g(X) = (f \otimes h)(X) = \int_{-\infty}^{\infty} f(x)h(X-x)\mathrm{d}x \tag{5.8}$$

To put in words: the image is the convolution of the object with the PSF.

5.5 Graphic method for the solution of the convolution integral

As any 1D integral represents the area below the integrand, $\int_{-\infty}^{\infty} f(x)h(X-x)\mathrm{d}x$ represents the area below the function $f(x)h(X-x)$. Calculating this area for every value of X we will have the convolution function. It is possible to calculate this area, following some steps, as will be shown.

---

[2] This notation prevents confusion between the convolution of two functions and their product. Additionally, it explicitly indicates that the convolution depends on X, not on x.

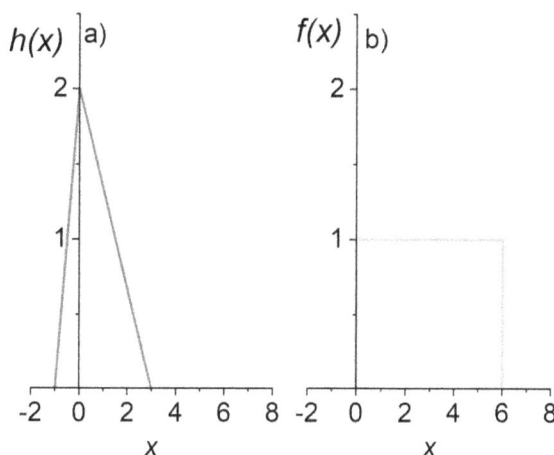

Figure 5.3. The functions $h(x)$ and $f(x)$.

Let us suppose that $h(x)$ has the form displayed in figure 5.3(a) (for more generality the function chosen as an example is not symmetric with respect to $x = 0$). This could be for example, the PSF for a given optical system. The function which describes the object is represented by $f(x)$ (figure 5.3(b)).

First step. Draw $h(-x)$ by mirroring $h(x)$ with respect to the $x = 0$ line as in figure 5.4(a). This function will be $h(X - x)$ with $X = 0$. It can be translated along the x-axis by changing the value of X. For positive values of X the function will be translated to positive values of the coordinate. Conversely, for negative values of X, the function will be translated to negative values of the coordinate. That is, by increasing/decreasing the X values the function shifts toward positive/negative values. To realize this, note, for example, that the maximum of the function occurs when the argument is zero and, as the argument is $X - x$, for $X = 1$ the maximum will be located at $X - x = 0$, or $x = 1$. For $X = -1$, the maximum will be located at $x = -1$.

Second step. The function will be translated up to a location on the x-axis in which there is just no overlap with $f(x)$. Figure 5.4(b) shows the $h(-x)$ together with $f(x)$ in the same graphics. As can be observed in this figure, for this value of X the functions overlap, then $h(-x)$ must be shifted to $X = -1$ as shown in figure 5.4(c). For $X < -1$ the function will not overlap, then the product $f(x)h(X - x)$, and the convolution integral will be zero.

Third step. The function is translated by changing the values of X. In every position, the area below the product of the functions $f(x)$ and $h(X - x)$ will be evaluated. This area will represent the value of the convolution $g(X)$. Such a translation process is shown in figures 5.4(d)–(m).

To illustrate the calculation of the area of the product $f(x) \, h(X - x)$, the case of $X = 1$ (figure 5.4(e)) is considered (see figure 5.5). For this case, $f(x)$ and $h(X - x)$ overlap for x ranging from 0 to 2 and the product of the functions is only different from zero in this region. As the function $f(x) = 1$, the product $f(x)h(X - x)$ will be

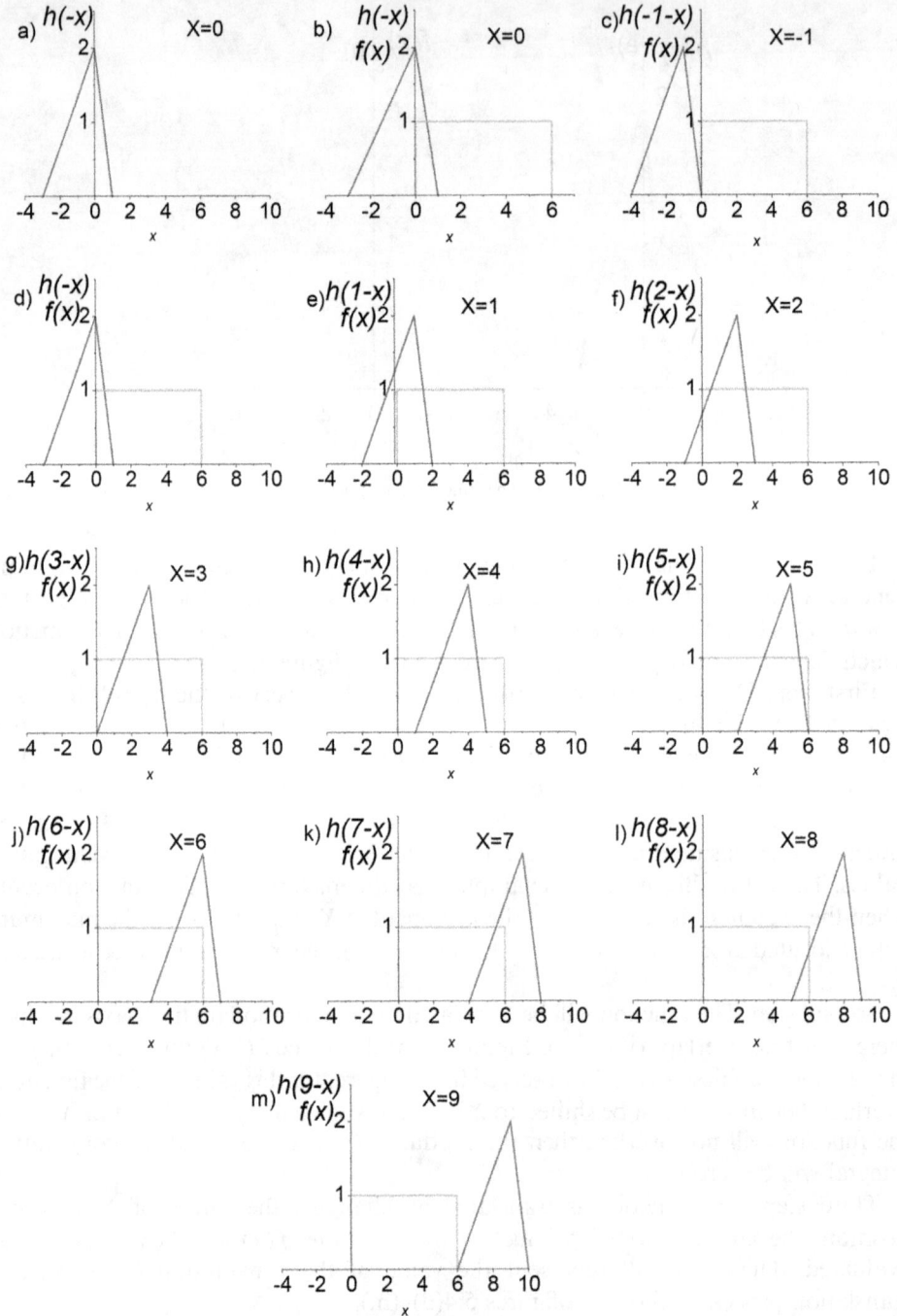

Figure 5.4. The graphic method for the convolution $(f \otimes h)(X)$. The function $h(-x)$ (a) is plotted together with the function $f(x)$ (b). (c) $h(-x)$ is shifted to the position immediately prior to overlapping ($X = -1$) (c). $h(-x)$ is translated across $f(x)$ (d–m) in steps of one. The area below the product $f(x) h(X - x)$ is calculated in every step.

equal to $h(X - x)$. Then, the area to be calculated is that highlighted with the blue lines pattern: the sum of the triangle DEF and the trapezium $CBDE$. The area of the triangle DEF can be calculated as $1/2(EF \times ED) = 1$. The area of the trapezium will be equal to the area of the triangle AED minus the area of the triangle ACB. To calculate this area, the value of the length CB is needed; it can be calculated using the relation $\frac{DE}{AE} = \frac{BC}{CA}$ which follows from the similarity of the triangles AED and ACB. From the data in the graphic, $BC = \frac{4}{3}$; and the area of the triangles ACB and AED are $\frac{4}{3}$ and 3, respectively. Then, the area highlighted in figure 5.5 is $\frac{8}{3} \approx 2.67$. With similar geometrical consideration, the areas for the other cases can be calculated. The results are displayed in the table below and figure 5.6.

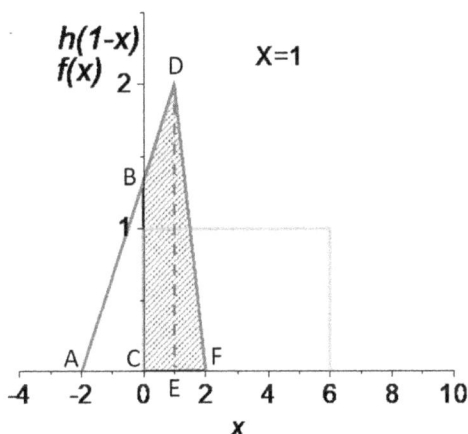

Figure 5.5. Graphic solution of the convolution value for $X = 1$. The area to be calculated is highlighted.

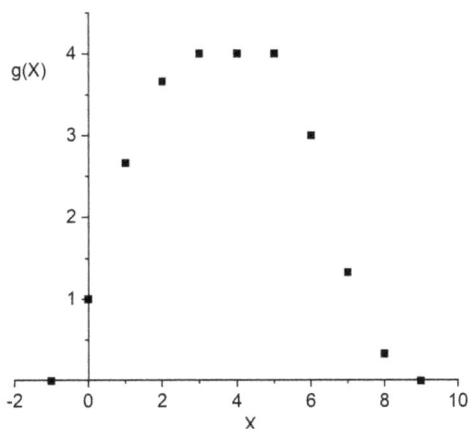

Figure 5.6. The convolution $(f \otimes h)(X)$ after the graphical solution.

X	-1	0	1	2	3	4	5	6	7	8	9
$g(X)$	0	1	8/3	11/3	4	4	4	3	4/3	1/3	0

As in the example above, whenever the functions being convolved are spatially restricted (i.e., functions with compact support), the domain of the resulting convolution function will equal the sum of the domains of the original functions.

5.6 Properties of the convolution

5.6.1 Commutativity

It is easy to demonstrate that the convolution operation is commutative. Let us consider the convolution integral:

$$(f \otimes h)(X) = \int_{-\infty}^{\infty} f(x)h(X - x)\mathrm{d}x$$

With the change of variable $x' = X - x$; $\mathrm{d}x' = -\mathrm{d}x$ and interchanging the integral limits due to the minus sign in the differential:

$$(f \otimes h)(X) = \int_{\infty}^{-\infty} f(X - x')h(x')(-\mathrm{d}x') = -\int_{\infty}^{-\infty} f(X - x')h(x')(\mathrm{d}x')$$

$$= \int_{-\infty}^{\infty} f(X - x')h(x')(\mathrm{d}x') = \int_{-\infty}^{\infty} h(x')f(X - x')(\mathrm{d}x') = \int_{-\infty}^{\infty} h(x)f(X - x)(\mathrm{d}x)$$

Since x' is a dummy variable.
   Then:

$$(f \otimes h)(X) = (h \otimes f)(X) \tag{5.9}$$

5.6.2 The convolution of a function with the delta function

Let us consider the convolution integral:

$$(f \otimes h)(X) = \int_{-\infty}^{\infty} f(x)h(X - x)\mathrm{d}x$$

And in the case in which $f(x) = \delta(x)$.

$$(\delta \otimes h)(X) = \int_{-\infty}^{\infty} \delta(x)h(X - x)\mathrm{d}x$$

And, according to the sifting property of the delta function ($\int_{-\infty}^{\infty} f(x)\delta(x)\mathrm{d}x = f(0)$, equation (3.17)):

$$(\delta \otimes h)(X) = \int_{-\infty}^{\infty} \delta(x)h(X - x)\mathrm{d}x = h(X - 0) \equiv h(X) \tag{5.10}$$

This means that the convolution of a function and the delta function is the function itself. Also, the convolution for a delta function shifted from zero $\delta(x - x_0)$ will be:

$$h(X - x_0) \tag{5.11}$$

So, the convolution of a function with the delta function translates the function to the coordinate of the delta. Of course, the convolution of a function with a series of delta functions, reproduces the function at the position of each delta function.

5.6.3 The convolution theorem

This is an important theorem which states that the convolution of two functions is equal to the product of their Fourier transforms, i.e.:

$$F\{(f \otimes h)(X)\} = F\{f(x)\}F\{h(x)\} \tag{5.12}$$

In fact:

$$F\{(f \otimes h)(X)\} = \int_{-\infty}^{\infty} \left[\int_{-\infty}^{\infty} f(x)h(X - x)dx\right] \exp(-i2\pi\xi X)dX$$

$$= \int_{-\infty}^{\infty} \left[\int_{-\infty}^{\infty} h(X - x)\exp(-i2\pi\xi X)dX\right] f(x)dx$$

Using the change of variable $x' = X - x$; $dx' = dX$

$$= \int_{-\infty}^{\infty} \left[\int_{-\infty}^{\infty} h(x')\exp(-i2\pi\xi(x' + x))dx'\right] f(x)dx$$

$$= \int_{-\infty}^{\infty} \left[\int_{-\infty}^{\infty} h(x')\exp(-i2\pi\xi x')\exp(-i2\pi\xi x)dx'\right] f(x)dx$$

$$= \int_{-\infty}^{\infty} \left[\int_{-\infty}^{\infty} h(x')\exp(-i2\pi\xi x')dx'\right] f(x)\exp(-i2\pi\xi x)dx$$

The expression between brackets can be taken out of the integral in x. Then:

$$= \int_{-\infty}^{\infty} h(x')\exp(-i2\pi\xi x')dx' \int_{-\infty}^{\infty} f(x)\exp(-i2\pi\xi x)dx$$

That is: $F\{f(x)\}F\{h(x)\}$

This theorem is quite useful. In some cases, it is more difficult to calculate the convolution of two functions than their Fourier transform. In these cases, it is better to first calculate the Fourier transform of the functions, multiply them (multiplication is a rather simple operation) and then apply the inverse Fourier transform to the result to obtain the convolution function. As an example, we can consider that the convolution of two square pulses is a triangular pulse (see problem 5.1). As the Fourier transform of each square pulse is the sinc function, it is expected, according to the convolution theorem that the Fourier transform of that convolution will be the sinc² function. This is in fact the result obtained in problem 3.3.

A corollary of the convolution theorem states that the Fourier transform of the product of two functions is equal to the convolution of their individual Fourier transforms. That is:

$$F\{fh\} = F\{f\} \otimes F\{h\} \tag{5.13}$$

As an example for this corollary, we can consider the cos $(2\pi\xi_0 x)$ function limited in space. Such a problem was proposed in chapter 3 (problem 3.7). This function will be the product of the cos $(2\pi\xi_0 x)$ function by the square pulse in such a way that the square pulse width limits the range of the cosine function. Then, according to the corollary of the convolution theorem, the Fourier transform of this product will be the convolution of their individual Fourier transforms. As the Fourier transform of the unlimited cos $(2\pi\xi_0 x)$ is a couple of delta functions located at ξ_0 and $-\xi_0$ (see equation (3.25)), and the Fourier transform of the square pulse is the sinc function, the resulting Fourier transform of the product will be two sinc functions centred at ξ_0 and $-\xi_0$ as could have been obtained by calculating directly the Fourier transform using equation (3.10).

5.7 Convolution in 2D

The convolution equation for the image formation $G(X) = f(x) \otimes h(x) = \int_{-\infty}^{\infty} f(x)h(X-x)\mathrm{d}x$ can be easily extended to 2D and results in:

$$G(X, Y) = (f \otimes h)(X, Y) = \int_{-\infty}^{\infty} f(x, y)h(X-x, Y-y)\mathrm{d}x\mathrm{d}y$$

Since the function $f(x, y)$ is a 2D function and represents a surface, the integral corresponds to the volume (not the area) under the product of the functions $f(x, y)$ and $h(X-x, Y-y)$. Taking this in consideration, a graphic method can be used for calculating the convolution as in the 1D case. To exemplify, let us consider the convolution of the 2D squared pulse with itself.

Figure 5.7 presents the contour maps of $f(x, y)$ and $h(x, y)$. This function can be written analytically as:

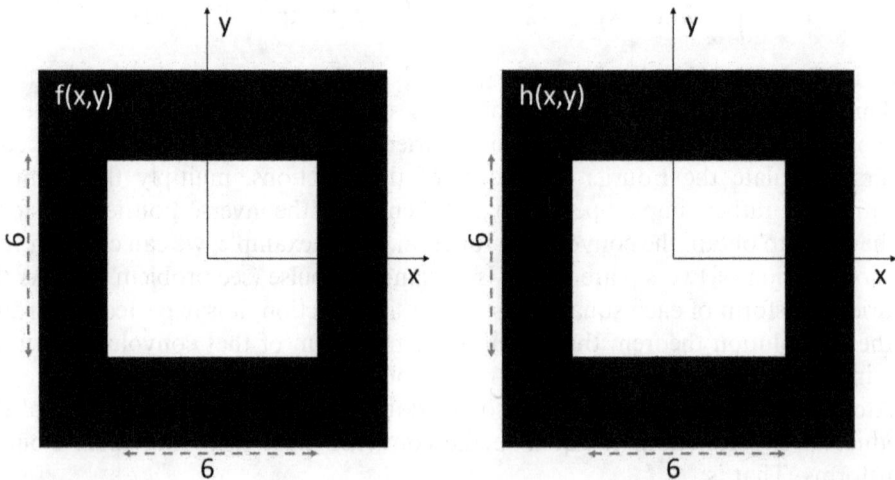

Figure 5.7. Contour maps of the 2D functions $f(x, y)$ and $h(x, y)$. In the contour, white represents 1 and black zero.

$$f(x, y) = h(x, y) = 1 \text{ for } |x, y| < 3 \text{ and } 0 \text{ in the rest of the space.}$$

In the contour map this is represented with two shades: white for 1 and black for 0.

In the first step, the function $h(x, y)$ is mirrored with respect to the lines $x = 0$ and $y = 0$ to become $h(-x, -y)$. As $h(x, y)$ is an even function in this case, the operation does not modify the function. In the second step $h(-x, -y)$ is translated along both axes X, Y up to a location in which there is just no overlap with $f(x, y)$. Figure 5.8(a) presents such a condition: the $h(-x, -y)$ has been translated to $X = -6$ and $Y = 6$ and now it is $h(-6 - x, 6 - y)$, for any value with $X \leqslant -6$ and $Y \geqslant 6$, the functions do not overlap (the separation between the major ticks on the scales is 1 and only the contours delimiting the areas where the functions are equal to 1 from the areas where they are equal to zero are shown in the figure. Then function $h(-x, -y)$ must be translated, step by step, over the whole area of the function $f(x, y)$, and in every step, the volume under the integral, that is, the convolution, is calculated. figure 5.8(b) shows a first step in which $h(-x, -y)$ has been translated to $X = -5$ and $Y = 5$ to become $h(-5 - x, 5 - y)$. Since the value of both functions is 1 on the interval in which they overlap, the product $f(x, y)h(-5 - x, 5 - y)$ in this interval will be equal to 1 as well (this will be the same for all the steps). The volume under the 2D function (the convolution) will be equal to the product of the

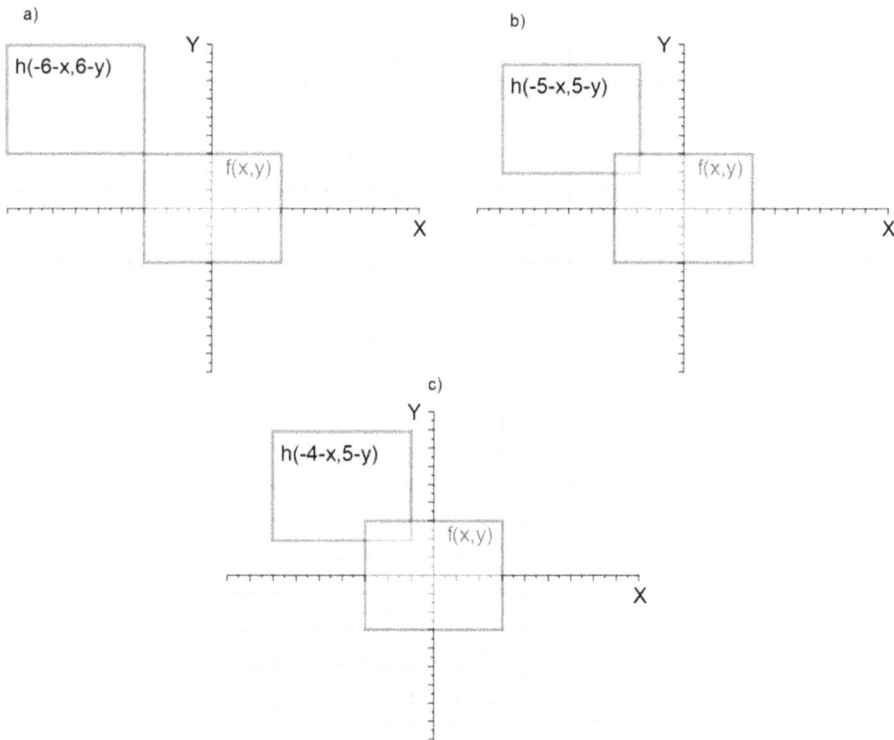

Figure 5.8. Illustration of the graphic method for the convolution of 2D functions. (a,b,c) the first step of the convolution according the text.

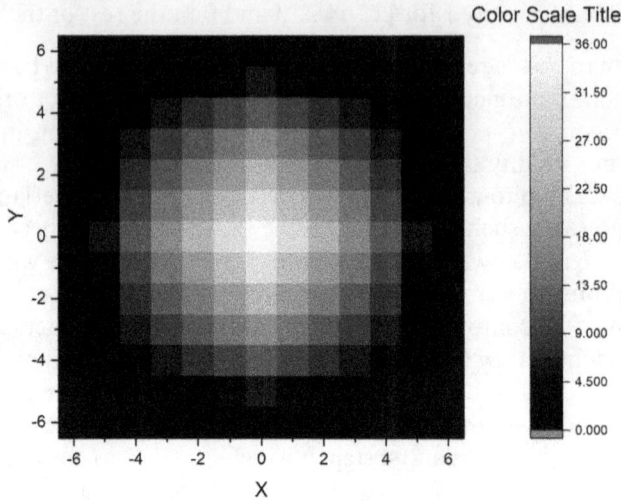

Figure 5.9. Grey shades contour map for the convolution of a 2D square pulse and itself.

functions (equal to 1) multiplied by the area of overlapping that in this step, is also 1. That means that the convolution for the coordinate $(-5, 5)$ is 1. In the second step, shown in figure 5.8(c), $h(-x, -y)$ will continue its trip along the row $Y = 5$, changing to $X = -4$. In this case the area (and the convolution) is 2.

The reader can continue the process and scan all the surface, calculating, in the same way, the convolution in every step. In the end, the matrix shown in table 5.1 is obtained.

From this matrix we can obtain 2D plots of the convolution result. In figure 5.9 a contour map in grey shades of the data is presented.

As expected, the maximum value of the convolution function occurs for $h(0, 0)$, when the two functions completely overlap. Also, it can be observed that the size of the region (in each direction) where the convolution is different from zero is double that of the same region in each function. In fact, in this function, this range is 6 in each direction, while in the convolution it is 12. The stepped character that is observed in the contour map comes from the scarcity of data taken (we calculated the convolution using steps of dimension unity along both axis). In real images the quality of the convolution will depend on the resolution of the image. However, generally, graphing software allows one to expand a matrix by incorporating not calculated (or measured) data through an interpolation process. For example, in the contour map of figure 5.10(a) the 13×13 matrix was expanded to a 39×39 matrix using the same data, and the stepping is less pronounced. Figure 5.10(b) shows a 3D colour surface plot from the initial data using interpolation. It is worth noting that this kind of interpolation could introduce artifacts, and it is always better to use the appropriate resolution in the convolution calculation

Following a similar procedure as explained above, the convolution of any 2D image can be calculated. In figure 5.11, the convolution of a circle function with itself is shown.

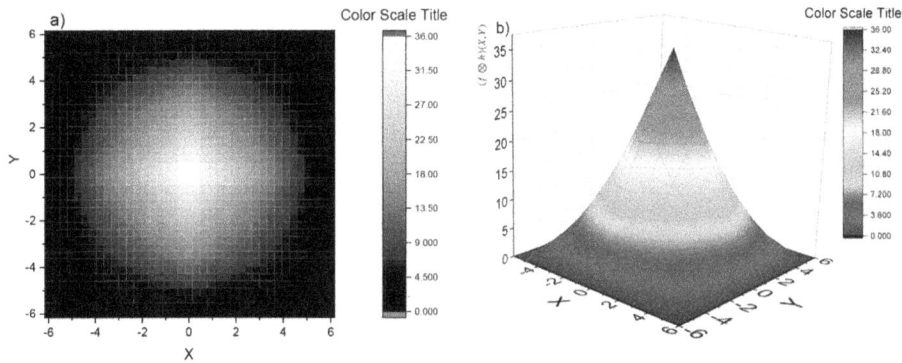

Figure 5.10. (a) Contour map using an expanded matrix of the convolution analysed in this paragraph. (b) 3D colour surface plot from the data of matrix of table 5.1 using interpolations.

Table 5.1. Matrix (13×13) of the convolution of the square pulse with itself.

↓Y X→	−6	−5	−4	−3	−2	−1	0	1	2	3	4	5	6
6	0	0	0	0	0	0	0	0	0	0	0	0	0
5	0	1	2	3	4	5	6	5	4	3	2	1	0
4	0	2	4	6	8	10	12	10	8	6	4	2	0
3	0	3	6	9	12	15	18	15	12	9	6	3	0
2	0	4	8	12	16	20	24	20	16	12	8	4	0
1	0	5	10	15	20	25	30	25	20	15	10	5	0
0	0	6	12	18	24	30	36	30	24	18	12	6	0
1	0	5	10	15	20	25	30	25	20	15	10	5	0
2	0	4	8	12	16	20	24	20	16	12	8	4	0
3	0	3	6	9	12	15	18	15	12	9	6	3	0
4	0	2	4	6	8	10	12	10	8	6	4	2	0
5	0	1	2	3	4	5	6	5	4	3	2	1	0
6	0	0	0	0	0	0	0	0	0	0	0	0	0

We have used the simple example of the convolution of a 2D square pulse with itself to illustrate the graphic method for calculating the convolution. It is interesting to note that, if $f(x, y)$ had represented the illumination of the object and $h(x, y)$ the PSF of the optical system, such an optical system would be a very inconvenient one. In fact, the obtained image (figures 5.9 or 5.10) would be very different from the object (figure 5.7(a)) as a consequence of the PSF and the object having comparable sizes. In general, a practical optical system requires that the size of the PSF be as small as possible, and at least smaller than the object.

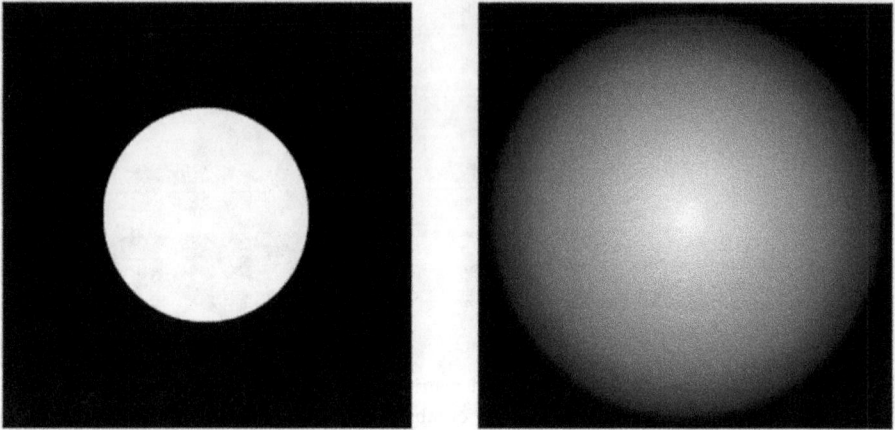

Figure 5.11. The contour map of a circle function (left), its convolution (right).

5.8 Problems

Problem 5.1 Demonstrate that the convolution of a square pulse (height H and width a) with itself is a triangular pulse. Make a sketch of the result showing the height and width of the resulting triangular pulse. Use the convolution theorem to calculate the Fourier transform of the triangular pulse.

Problem 5.2 Using the graphic method (figure 5.12), determine the convolution $(f \otimes h)(X)$.

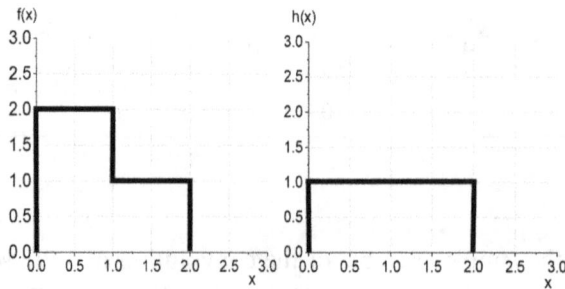

Figure 5.12.

Problem 5.3 Draw the convolution $\delta(x - 2)$ with $f(x)$ in the empty frame (figure 5.13).

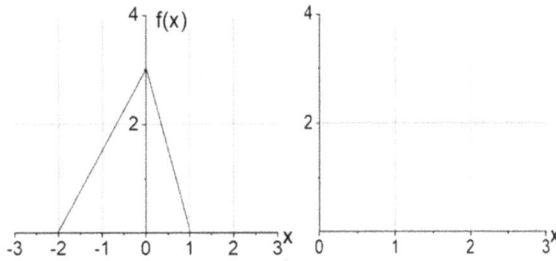

Figure 5.13.

Problem 5.4 Consider the series $f(x) = \sum\limits_{n=-3}^{2} \delta(x - 2n)$ where n is an integer.

(a) Graphically represent the function $f(x)$.
(b) Make the convolution of $f(x)$ with the square pulse of width 1.5 and represent it graphically.
(c) Derive an expression for the Fourier transform of this convolution. Describe the obtained function.

Further Reading

[1] Goodman J W 2017 *Introduction to Fourier Optics* 4th edn (New York: W.H. Freeman)
[2] Gaskill J D 1978 *Linear Systems, Fourier Transforms, and Optics* (New York: Wiley)
[3] Hespanha J P 2018 *Linear Systems Theory* 2nd edn (Princeton, NJ: Princeton University Press)
[4] Oppenheim A V and Willsky A S 1996 *Signals and Systems* 2nd edn (Upper Saddle River, NJ: Prentice-Hall)
[5] Khare K 2015 *Fourier Optics and Computational Imaging* (Hoboken, NJ: Wiley)

Chapter 6

Fraunhofer diffraction and Fourier transform

This chapter presents the principles of diffraction within the framework of Fourier optics, offering a comprehensive understanding of how diffraction phenomena can be analysed and interpreted through Fourier transforms using the aperture function and the relation between the spatial frequency and the position on the screen. Various examples are presented in which different diffraction objects are studied using these concepts, including single and multiple slits, the diffraction grating and the bidimensional square aperture. The influence of considering the finite width of the slits is analysed as well. The apodization technique is presented as an example of the application of the present concepts for enhancing the capacities of optical systems.

6.1 Fraunhofer diffraction. Aperture function

One important application of the Fourier transform formalism is the analysis of diffraction. As will be seen in the present chapter, it allows one to study almost all diffraction cases with the same algorithm, using the concept of the aperture function. To illustrate the relation between the diffraction and the Fourier transform, let us take into consideration the Fraunhofer approximation (equation (2.10)) with the obliquity factor equal to one, and making $dS = dx\,dy$, and $\overrightarrow{OP} = x\overrightarrow{u_x} + y\overrightarrow{u_y}$. That is:

$$\mathcal{E}(M) = \frac{E_o \exp(ik\text{OM})}{\text{OM}} \iint_{(x,y)\,\in\text{hole}} dx\,dy \exp\left(\frac{-ik\overrightarrow{\text{OM}}.(x\overrightarrow{u_x} + y\overrightarrow{u_y})}{\text{OM}}\right) \quad (6.1)$$

where x,y are the coordinates in the object plane; $\overrightarrow{u_x}$, $\overrightarrow{u_y}$ the unitary vectors in the x and y directions and the double integral run over the hole area. Other symbols have the same meaning as in section 2.5. The double integral runs along the object plane. Using the angles θ_x and θ_y, as defined in figure 2.22, equation (6.1) becomes (see section 2.5.4.3):

doi:10.1088/978-0-7503-6392-1ch6

$$\underline{\mathcal{E}}(M) = \frac{E_o \exp{(ik\text{OM})}}{\text{OM}} \iint \exp{\left\{-ik[x\sin{(\theta_x)} + y\sin{(\theta_y)}]\right\}}dx\,dy \qquad (6.2)$$

In section 2.5.1, we assumed that the incident field maximum amplitude was uniform over the object plane, and equal to a constant E_o. However, if the object plane is not uniformly illuminated, the incident field maximum amplitude depends on x and y, and we have to write it $E_o(x, y)$ and include it in the integral. Also, the object plane can contain the mask, which lets the light through, or partially, or not at all, depending on the position (x,y) where the field crosses the mask. Mathematically, this corresponds to the multiplication of the incident field by what we call the transmission function of the mask. This function $T(x, y)$ is equal to unity (respectively zero) for points belonging to a transparent (respectively opaque) part of the mask, and a value between 0 and 1 for partially transparent regions.

The mask can also impose phase variations to the field crossing it, and this extra phase can also vary with x and y. Thus, the field leaving the object plane is equal to the incident field multiplied not only by the transmission $T(x, y)$ but also by an extra term: $\exp(i\varphi_{\text{mask}}(x, y))$. That leads us to the following equation:

$$\underline{\mathcal{E}}(M) = \frac{\exp{(ik\text{OM})}}{\text{OM}} \int\!\!\int_{-\infty}^{\infty} E_o(x, y)T(x, y)\exp(i\varphi_{\text{mask}}(x, y)) \cdot \exp{\left\{-ik[x\sin{(\theta_x)} + y\sin{(\theta_y)}]\right\}}dxdy$$

Note that, instead of integrating solely over the transmitting area of the mask, we have now extended the integration to the entire (x,y) plane. This is possible because the integrand vanishes outside the aperture, where the transmission function $T(x, y) = 0$.

The intensity at any point M of the image plane will be calculated by multiplying $\underline{\mathcal{E}}(M)$ by its complex conjugate. In any case then, the term $\exp{(ik\text{OM})}$ will be multiplied by $\exp{(-ik\text{OM})}$ and then leads to a factor 1 in the intensity. Furthermore, let us recall that we are under the paraxial approximation (see paragraph 2.5.3.2), which means, among other things, that $X, Y \ll Z$ where $(X, Y, Z = D)$ are the coordinates of point M. That means that $\frac{1}{\text{OM}}$ does not vary much with the position of M on the screen. That means we can simply forget about the dependency of $\frac{\exp{(ik\text{OM})}}{\text{OM}}$ on the position of M.

Let us define the aperture function $A(x, y)$ so that:

$$\underline{\mathcal{E}}(M) = \int\!\!\int_{-\infty}^{\infty} A(x, y)\exp{\left\{-ik[x\sin{(\theta_x)} + y\sin{(\theta_y)}]\right\}}dxdy \qquad (6.3)$$

where:

$$A(x, y) = \frac{\exp{(ik\text{OM})}}{\text{OM}}E_o(x, y)T(x, y)\exp(i\varphi_{\text{mask}}(x, y))$$

In a nutshell, the aperture function is a complex valued function that accounts for the spatial dependence of $E_o(x, y)$ (translating a non-uniformity of the illumination), the transmission of the mask (that also depends on x and y), and any possible additional phase in the object plane.

The k vector is normal to the equal phase-lines of the wave, and it is colinear with OM, then, in equation (6.2), $k \sin (\theta_x) = k_x$ and $k \sin (\theta_y) = k_y$. On the other hand, $k_x = 2\pi\xi$, and $k_y = 2\pi\eta$. Then:

$$\mathcal{E}(M) = \iint_{-\infty}^{\infty} A(x, y) \exp [-i(k_x x + k_y y)] dx \, dy$$

Or:

$$\mathcal{E}(M) = \iint_{-\infty}^{\infty} A(x, y) \exp [-i(2\pi\xi x + 2\pi\eta y)] dx \, dy \qquad (6.4)$$

The reader will note the similitude of this equation with equation (4.1) for the Fourier transform of a 2D function. According to it:

> The complex amplitude of the electric field in the Fraunhofer diffraction pattern is the Fourier transform of the aperture function.

In one dimension:

$$\mathcal{E}(M) = \int_{-\infty}^{\infty} A(x) \exp (-i2\pi\xi x) dx dy$$

It can be noted that $\mathcal{E}(M) = \mathcal{E}(\xi)$ since the position of M in the screen is determined by the value of ξ. In fact:

$$2\pi\xi \equiv k_x = k \sin (\theta_x) = \frac{2\pi}{\lambda} \sin (\theta_x) \approx \frac{2\pi}{\lambda} \tan (\theta_x) = \frac{2\pi}{\lambda} \frac{X}{D}$$

X being the coordinate of M in the screen. Then:

$$\xi = \frac{X}{\lambda D} \qquad (6.5)$$

Equation (6.5) relates the spatial frequency with the position of M in the screen. Then, we can write:

$$\mathcal{E}(\xi) = \int A(x) \exp (-i2\pi\xi x) dx \qquad (6.6)$$

In 2D:

$$\mathcal{E}(\xi, \eta) = \iint A(x, y) \exp [-i(2\pi\xi x + 2\pi\eta y)] dx \, dy \qquad (6.7)$$

For the 1D simple case, some equal-phase planes are shown in figure 6.1 for the light arriving to the point M. The k vector normal to the equal-phase lines and its components k_x and k_z are also displayed. The component k_x can be written as a function of the spatial frequency, ξ, along the x-axis as $2\pi\xi$.

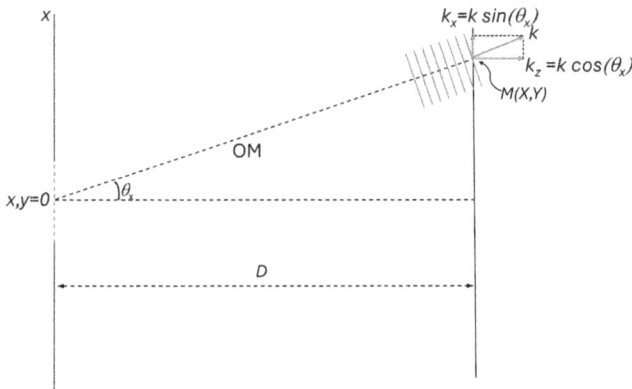

Figure 6.1. Equal-phase lines for the wave coming from the aperture in the θ_x direction. The wave vector and its components are displayed.

6.2 Examples. One, two, three, several slits

6.2.1 One uniformly illuminated slit of width a

As the first example of application of equation (6.7) let us consider a single slit of width a uniformly illuminated as in section 2.5.4.1. For this case the aperture function is:

$$A(x) = \begin{cases} A_0, & |x| < a/2 \\ 0, & |x| \geq a/2 \end{cases}$$

The slit and the aperture function are represented in figure 6.2.

$$\underline{\mathcal{E}}(\xi) = \int_{-a/2}^{a/2} A(x) \exp\left(-i2\pi\xi x\right)\mathrm{d}x$$

This problem was resolved in section 3.2.2 and the result is:

$$\underline{\mathcal{E}}(\xi) = A_0 a \, \text{sinc}(\pi\xi a) \tag{6.8}$$

This equation represents the dependence of the complex amplitude of the electric field on the spatial frequency ξ. As commented above, it can be noted that equation (6.8) gives also the dependence of the complex electric field amplitude with X, the position on the screen. According to equation (6.5):

$$X = \lambda D\xi \tag{6.9}$$

The zeros of the function appear when $\text{sinc}(\pi\xi a) = 0$, when the spatial frequency $\xi = \pm n\frac{1}{a}$ (with $n = 1,\ 2,\ 3,\ \ldots$) or when the coordinate $X = \pm n\frac{\lambda D}{a}$. The graphical representation of $\underline{\mathcal{E}}(\xi)$ and $\underline{\mathcal{E}}(X)$ are shown in figure 6.3.

The quantity which is measured in a diffraction pattern is the intensity (I) instead of the electric field amplitude. The intensity of the wave is proportional to the square of the modulus of the complex amplitude of the electric field (see equation (2.8));

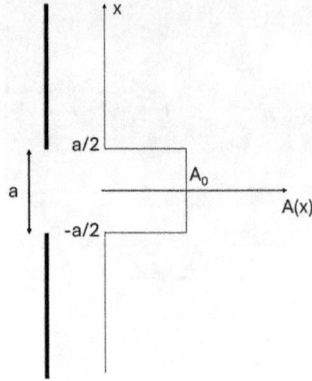

Figure 6.2. The aperture function for an illuminated single slit of width a.

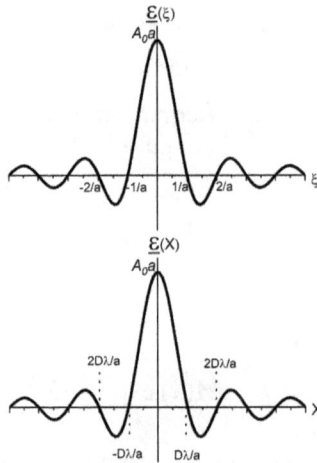

Figure 6.3. The distribution of the complex amplitude of the electric field as a function of the spatial frequency along the x-axis (top), or the position on the screen, X (bottom).

$I(\xi)$ and $I(X)$ are represented in figure 6.4 for the present case. As expected, they present the same shape as in figure 2.18.

6.2.2 Two slits of infinitesimal width. The Young experiment

The aperture function, $A(x)$ of a double slit, each with infinitesimal width, and separated by a distance b can be represented by a couple of delta functions located at the position of the slits. Let's say that the slits are located at $\frac{b}{2}$ and $-\frac{b}{2}$, then:

$$A(x) = A_0 \left[\delta\left(x - \frac{b}{2}\right) + \delta\left(x + \frac{b}{2}\right) \right]$$

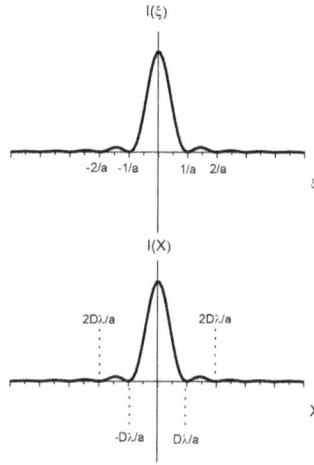

Figure 6.4. The distribution of the intensity as a function of the spatial frequency along the x-axis (top), or the position on the screen, X (bottom).

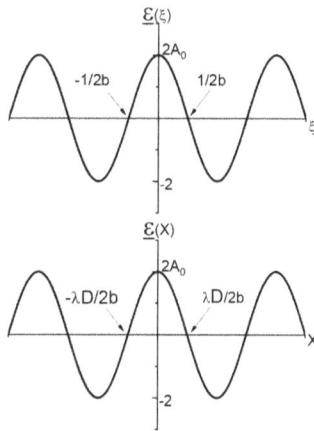

Figure 6.5. The distribution of the complex amplitude of the electric field as a function of the spatial frequency along the x-axis (top), or the position on the screen, X (bottom), for two slits of infinitesimal width.

And the electric field amplitude will be:

$$\mathcal{E}(\xi) = \int_{-\infty}^{\infty} A(x) \exp\left(-i2\pi\xi x\right) dx = \int_{-\infty}^{\infty} A_0 \left[\delta\left(x - \frac{b}{2}\right) + \delta\left(x + \frac{b}{2}\right) \right] \exp\left(-i2\pi\xi x\right) dx$$

Using the sifting property of the delta function and the Euler formula it is easy to find that:

$$\mathcal{E}(\xi) = 2A_0 \cos\left(\pi\xi b\right) \qquad (6.10)$$

Using equation (6.9), the first zeros of the functions $\mathcal{E}(\xi)$ and $\mathcal{E}(X)$ will be: $\xi = \frac{1}{2b}$ and $x = \frac{\lambda D}{2b}$, respectively as shown in figure 6.5. Using equation (6.9), the complex amplitude of the electric field can be expressed as a function of X as:

$$\mathcal{E}(X) = 2A_0 \cos \left(\pi \frac{X}{\lambda D} b \right) \tag{6.11}$$

The square of the modulus of these complex amplitude functions represents the intensity on the screen: a set of equally spaced strips, the famous result of the Young experiment as in section 2.4.7.

6.2.3 Two slits of finite width

Figure 6.6 represents the aperture function of a double slit, but unlike to the previous case, now the slits have a finite width a. It can be noted that this aperture function can be obtained by reproducing the same square pulse at two different coordinates: $x = \pm \frac{b}{2}$. From equations (5.10) and (5.11), such a function can be expressed as:

$$A(x) = \left[\delta\left(x - \frac{b}{2} \right) + \delta\left(x + \frac{b}{2} \right) \right] \otimes A_0 \text{Rect}_a(x)$$

The function $\text{Rect}_a(x)$ is the square pulse of width a as defined in equation (3.11).

Then, using the convolution theorem:

$$F\{A(x)\} = F\left\{ \delta\left(x - \frac{b}{2} \right) + \delta\left(x + \frac{b}{2} \right) \right\} A_0 F\{\text{Rect}_a(x)\}$$

Or:

$$F\{A(x)\} \equiv \mathcal{E}(\xi) = 2 \cos (\pi \xi b) A_o a \, \text{sinc}(\pi \xi a) \tag{6.12}$$

where we used result (3.12) and applied the property presented in section 3.3.4 (about the Fourier transform of the Fourier transform) to the Fourier transform of cos (see equation (3.25)). The obtained result is a cosine function modulated by a sine cardinal function as presented in figure 2.21.

6.2.4 Three slits

The aperture function of the case of three infinitesimal slits located at $x = 0$, $\pm \frac{b}{2}$ can be represented by:

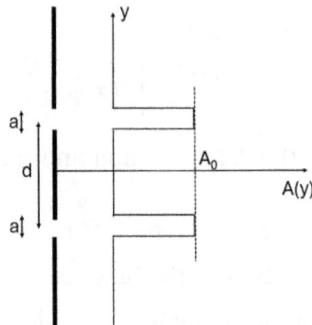

Figure 6.6. The aperture function for two illuminated slits of width a separated by a distance b.

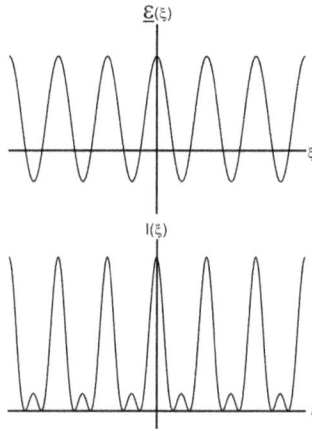

Figure 6.7. Complex amplitude of the electric field and Intensity as a function of the spatial frequency along x in the Fraunhofer diffraction pattern of the three infinitesimal slits.

$$A(x) = A_o \left[\delta(0) + \delta\left(x - \frac{b}{2}\right) + \delta\left(x + \frac{b}{2}\right) \right]$$

And the distribution of the complex amplitude of the electric field in the Fraunhofer approximation:

$$\mathcal{E}(\xi) = F \left\{ A_o \left[\delta(0) + \delta\left(x - \frac{b}{2}\right) + \delta\left(x + \frac{b}{2}\right) \right] \right\}$$

Which is:

$$\mathcal{E}(\xi) = A_o[1 + 2\cos(\pi\xi b)]$$

As in the other cases, the intensity will be proportional to the square of the modulus of the complex amplitude of the electric field.

$$I(\xi) \propto [1 + 2\cos(\pi\xi b)]^2$$

$\mathcal{E}(\xi)$ and $I(\xi)$ are represented in figure 6.7.

6.2.5 The diffraction grating

The diffraction grating formed by infinitesimal slits can be represented by a series of delta functions. If the number of slits is very large, the series can be approximated by the comb function (equation (3.20)):

$$A(x) = A_o \sum_{-\infty}^{\infty} \delta(x - nb)$$

in which b is the period of the diffraction grating.

6-8

And the distribution of the complex amplitude of the electric field in the Fraunhofer approximation will be also a comb in the frequency domain (equation (3.24)).

$$F\{A(x)\} \equiv \mathcal{E}(\xi) = \frac{A_o}{b} \sum_{-\infty}^{\infty} \delta(\xi - n\xi_0)$$

with $\xi_0 = 1/b$.

6.3 Diffraction in 2D

In the 2D case (equation (6.7)):

$$\mathcal{E}(\xi, \eta) = \int_{-\infty}^{\infty} \int_{-\infty}^{\infty} A(x, y) \exp(-i2\pi\xi x - i2\pi\eta y) \, \mathrm{d}x\mathrm{d}y$$

Let's consider the square aperture. In this case (see secton 4.2.1) the aperture function can be written as:

$$A(x, y) = g(x)h(y) \tag{6.13}$$

with $g(x) = 1$ for $|x| < \frac{a}{2}$; and $h(y) = 1$ for $|y| < \frac{a}{2}$ and 0 for $|x| > \frac{a}{2}$; $|y| > \frac{a}{2}$. So, the distribution of the complex amplitude of the electric field $\mathcal{E}(\xi, \eta)$ is given by:

$$F\{A(x, y)\} \equiv \mathcal{E}(\xi, \eta) = a^2 \, \mathrm{sinc}(\pi\xi a) \, \mathrm{sinc}(\pi\eta a)$$

The form of this function was shown in figure 4.8 in both a surface plot and in a contour map. The reader can reproduce this result from equation (2.15) considering that $a = b$ and that the factor $\frac{E_0 \exp(ik\mathrm{OM})}{\mathrm{OM}} = 1$.

6.4 Apodization

Coming from Greek 'feet removal', in diffraction, apodization names the process of reduction of the intensity of the lateral lobes of the diffraction pattern. These lateral lobes appear for example in the pattern of a finite slit (cardinal sin function) or a circular hole (Bessel function) and contribute to reducing the resolution of microscopes and telescopes. These lateral lobes correspond to relatively larger frequency harmonics needed to reproduce abrupt intensity changes in the object plane. They can be eliminated by modifying the aperture function in such a way that the abrupt changes in intensity are reduced.

For example, the objective lens of a microscope can be coated with a thin film which partially absorbs the light hitting the lens. To reduce the abrupt change in intensity occurring in the border of the lens, the thickness of the coating must be increased from the centre to the border of the lens. As the intensity of the light going through the coating will decrease with the thickness, the above procedure will result in an intensity that decreases monotonically from the centre of the lens towards the border, without abrupt changes. In the case that, using the non-uniform coating, a Gaussian shape of the illumination is produced, the distribution of the electric field of the diffraction pattern will be the Fourier transform of the radial Gaussian function. But, as demonstrated in section 3.5.2, the Fourier transform of the Gaussian function is a Gaussian function itself, and the Gaussian function has no lobes. The diffraction pattern of this apodized aperture will be a Gaussian function (no lobes) instead of a Bessel function.

6.5 The array theorem

The distribution of the complex amplitude of the electric field of an array of identical apertures in the Fraunhofer diffraction is equal to the Fourier transform of an individual aperture multiplied by the Fourier transform of an array of delta functions (point sources) arranged in the same configuration of the apertures array.

This theorem can be easily demonstrated by considering that an array of N identical apertures in one dimension can be written as (see section 5.6.2):

$$A(x) = A_I(x) \otimes \sum_{i=1}^{N} \delta(x - x_{0i})$$

In which x_{0i} is the position of the ith aperture. Using the convolution theorem (equation (5.12)):

$$F\{A(x)\} = F\{A_I(x)\} \cdot F\left\{\sum_{i=1}^{N} \delta(x - x_{0i})\right\}$$

which is the mathematical expression of the array theorem.

6.6 Problems

Problem 6.1
(a) Using a graphing software plot equation (6.12) for different values of a and b, and calculate the first minimum in every case. Consider the case in which $a \ll b$ and compare with the result in equation (6.10).
(b) Write the corresponding equation $\underline{\mathcal{E}}(\xi)$.
(c) Demonstrate that the intensity can be expressed as in section 2.5.4.2.
(d) Represent the graphic of the intensity as a function of X.

Problem 6.2 How the diffraction pattern shown in figure 6.7 would be modified if the slits were finite instead of infinitesimal?

Problem 6.3 How the diffraction pattern of a diffraction grating would be affected if (a) the width of the grating slits is finite; (b) the number of slits is finite? (See problem 2.6.2.)

Problem 6.4 Build a digital image of:
  (a) a white square (black outside).
  (b) a white circle (black outside).

Find the intensity of the Fraunhofer Diffraction pattern using a Fourier transform software. Compare with the results of section 2.5.4.4.

Problem 6.5 Use the array theorem to find the distribution of electric field in the Fraunhofer diffraction pattern of the double slit of finite width.

Further Reading

[1] Goodman J W 2017 *Introduction to Fourier Optics* 4th edn (New York: W.H. Freeman)
[2] Born M and Wolf E 1999 *Principles of Optics* 7th edn (Cambridge: Cambridge University Press)
[3] Hecht E 2017 *Optics* 5th edn (Boston, MA: Pearson)
[4] Khare K 2015 *Fourier Optics and Computational Imaging* (Hoboken, NJ: Wiley)
[5] Saleh B E A and Teich M C 2019 *Fundamentals of Photonics* 3rd edn (Hoboken, NJ: Wiley)

IOP Publishing

Elementary Fourier Optics for Science and Engineering Students

Hélène Ollivier and Osvaldo de Melo

Chapter 7

Correlations. Transfer functions and images formation

This chapter defines and explores the properties of cross-correlations and autocorrelations, essential tools in image processing and analysis. Parseval's theorem, which connects the spatial and frequency domains, is derived, offering insights into energy distribution across these domains. Using the definitions of autocorrelation and Parseval's theorem, the Wiener–Khinchin theorem is presented as an application. Furthermore, the point spread function (PSF), and the convolution theorem are used to define the optical transfer function (OTF) as a complex quantity that includes the modulation transfer function (MTF) and the phase transfer function (PTF). The chapter examines the dependence of the OTF on frequency. Additionally, the role of lenses as Fourier transformers is discussed, along with the 4F imaging system, which exemplifies how optical systems can process and manipulate images effectively.

7.1 Cross-correlation. Examples

Suppose we need to locate a specific pattern within a larger image—for instance, identifying a particular house roof on a city map or detecting a bacterium in a blood sample image. To achieve this, the pattern must be shifted across the entire image until a match is found. However, if the image is large or the pattern appears in multiple locations, performing this task 'by eye' can quickly become cumbersome, time-consuming, or even impractical. Fortunately, when digital images of both the pattern and the larger image are available, a mathematical operation can simplify the process: cross-correlation.

For a 1D case, for real values functions, this operation is defined as:

$$(f \circledast h)(X) = \int_{-\infty}^{\infty} f(x)h(x - X)\mathrm{d}x \qquad (7.1)$$

and for 2D:

$$(f \circledast h)(X, Y) = \int_{-\infty}^{\infty} f(x, y)h(x - X, y - Y)dxdy \qquad (7.2)$$

In these equations, $h(x)$ and $f(x)$ (or $h(x, y)$ and $f(x, y)$ in 2D) represent the 'pattern' we want to compare and the 'larger function', respectively, as described in the examples above.

Note that cross-correlation is similar to convolution, with the key difference being the sign of the argument $x - X$ (or $x - X$, $y - Y$ in 2D), which is opposite in cross-correlation compared to convolution. A graphical method can be used to evaluate the integral, as in the case of convolution, but skipping the step in which the function $h(x)$ is mirrored. The variable X (X and Y in 2D) is used to shift the pattern across the larger image. Note that if X, $Y > 0$ the pattern is shifted toward positive values of x,y.

As an example, consider the functions $f(x)$ and $h(x)$, illustrated in figure 7.1. To compute the cross-correlation, the function $h(x)$ is shifted across $f(x)$. This is accomplished by constructing the shifted function $h(x - X)$, then incrementally varying X step by step. Let's start at $X = 0$, with $h(x - 0) \equiv h(x)$, where the functions do not overlap.

To determine the cross-correlation function, one must calculate the area below the product of the functions in every step. The resulting cross-correlation function, calculated with steps of 0.05, is shown in figure 7.2.

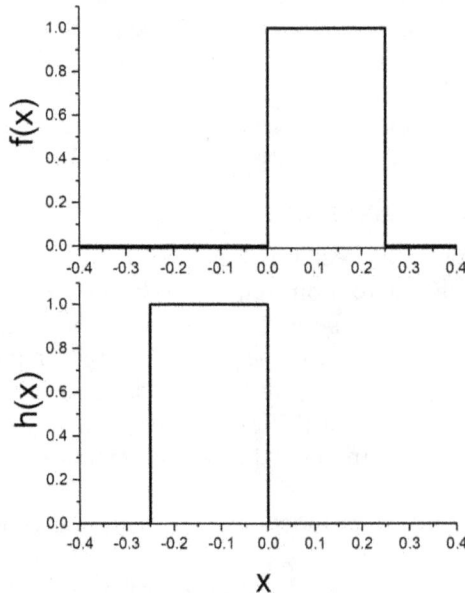

Figure 7.1. Functions $f(x)$ and $h(x)$.

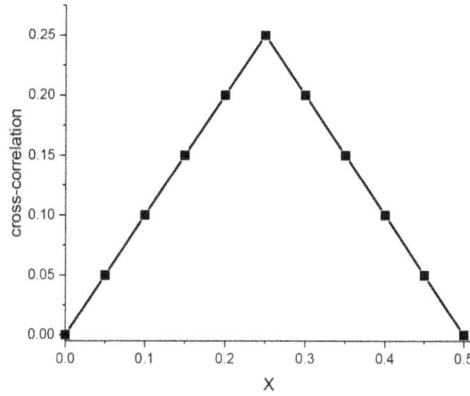

Figure 7.2. Cross-correlation of functions $f(x)$ and $h(x)$. Steps of 0.05 were used to obtain the square dots in the graph.

It is evident that the cross-correlation function reaches its maximum value when the two functions are perfectly aligned. An animated convolution is shown in figure 7.3.

7.2 Properties of the cross-correlation

7.2.1 Commutativity

It is straightforward to demonstrate that, unlike to convolution, cross-correlation, for real valued functions is not commutative in general. Instead:

$$(f \circledast h)(X) = \int_{-\infty}^{\infty} f(x)h(x-X)dx = \int_{-\infty}^{\infty} f(x+X)h(x)dx \neq \int_{-\infty}^{\infty} f(x-X)h(x)dx$$

To show this, we have done the following change of variable: $u = x - X$, so $dx = du$. So we have:

$$\int_{-\infty}^{\infty} f(x)h(x-X)dx = \int_{-\infty}^{\infty} f(u+X)h(u)du$$

Since u is a dummy variable, we can write this as:

$$\int_{-\infty}^{\infty} f(u+X)h(u)du \equiv \int_{-\infty}^{\infty} f(x+X)h(x)dx$$

This result clearly shows that cross-correlation is not commutative because changing the order of $f(x)$ and $h(x)$ yields a different expression.

It can also be observed that, if $h(x)$ is an even function, then, $h(x-X) = h(X-x)$. In this case, the distinction between convolution and cross-correlation disappears. The reason is that the mirroring operation inherent to convolution becomes redundant given the symmetry of $h(x)$. Specifically:

$$(f \circledast h)(X) = \int_{-\infty}^{\infty} f(x)h(x-X)dx = \int_{-\infty}^{\infty} f(x)h(X-x)dx = (f \otimes h)(X)$$

Where $(f \otimes h)(X)$ denotes the convolution of $f(x)$ and $h(x)$ (see chapter 5).

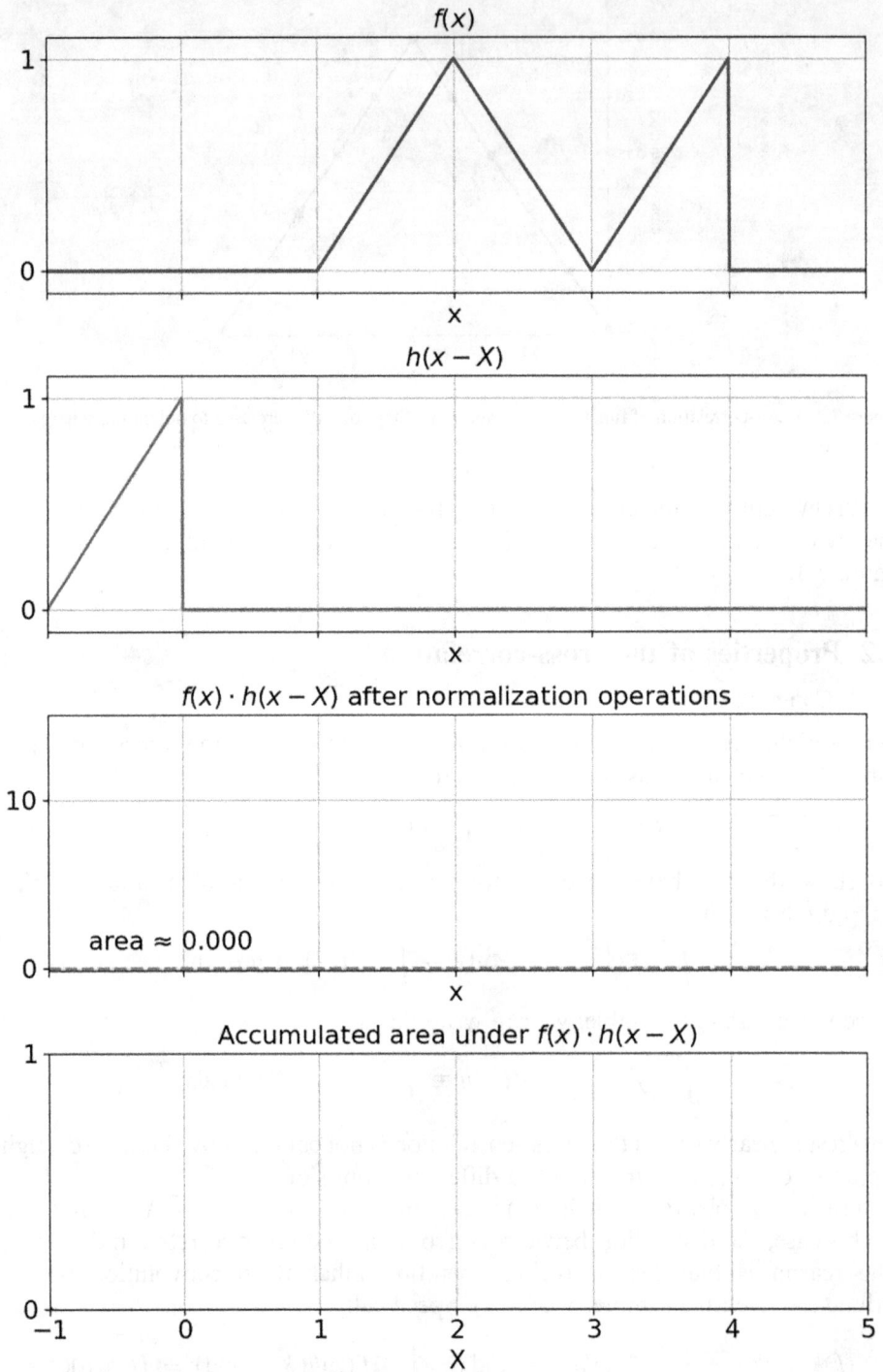

Figure 7.3. An animation of the correlation of two functions. There is an animated version of this figure, which is available online at https://doi.org/10.1088/978-0-7503-6392-1.

7.2.2 Autocorrelation

The autocorrelation is the correlation of a function with itself. If the function is real, it is defined as:

$$(f \circledast f)(X) = \int_{-\infty}^{\infty} f(x)f(x - X)\mathrm{d}x \tag{7.3}$$

The autocorrelation function reveals patterns or periodicities within the function. This is because, if there is periodicity, the autocorrelation will show multiple maxima in addition to the maximum at $X = 0$. For example, for the periodic function $f(x) = \sin(2\pi\xi_o x)$, the autocorrelation will present peaks at $x = 0$, $\pm 1/\xi_o$, $\pm 2/\xi_o$... This is often used in signal processing, optics, and time series analysis.

A different notation is sometimes used in which the cross-correlation of the function $f(x)$ and $h(x)$ in called C_{fh} and the autocorrelation of $f(x)$, C_{ff}.

7.3 The Parseval's theorem and the autocorrelation

7.3.1 The Parseval's theorem

Parseval's theorem states that 'the total energy of a signal in the spatial domain is equal to the total energy of its representation in the frequency domain'. To demonstrate this theorem, let us consider the complex function, $f(x)$. The squared magnitude of this function is:

$$|f(x)|^2 = f(x) \cdot f^*(x)$$

where $f^*(x)$ denotes the complex conjugate of $f(x)$. This can be also expressed as:

$$|f(x)|^2 = f(x) \cdot [F^{-1}\{F(\xi)\}]^*$$

since $f(x) = F^{-1}\{F(\xi)\}$. Integrating it from $-\infty$ to $+\infty$:

$$\int_{-\infty}^{\infty} |f(x)|^2 \mathrm{d}x = \int_{-\infty}^{\infty} f(x) \cdot \left[\int_{-\infty}^{\infty} F(\xi)\exp(i2\pi\xi x)\mathrm{d}\xi\right]^* \mathrm{d}x$$

Using the property $(F(\xi)\exp(i2\pi\xi x))^* = F(\xi)^*\exp(-i2\pi\xi x)$, the equation becomes:

$$\int_{-\infty}^{\infty} |f(x)|^2 \mathrm{d}x = \int_{-\infty}^{\infty} f(x) \cdot \left[\int_{-\infty}^{\infty} F(\xi)^*\exp(-i2\pi\xi x)\mathrm{d}\xi\right]\mathrm{d}x \tag{7.4}$$

This expression can be rearranged as:

$$\int_{-\infty}^{\infty} |f(x)|^2 \mathrm{d}x = \int_{-\infty}^{\infty} F(\xi)^* \left[\int_{-\infty}^{\infty} f(x)\exp(-i2\pi\xi x)\mathrm{d}x\right]\mathrm{d}\xi \tag{7.5}$$

The term in the square brackets is the Fourier transform of $f(x)$ ($F(\xi)$), and equation (7.5) simplifies to:

$$\int_{-\infty}^{\infty} |f(x)|^2 \mathrm{d}x = \int_{-\infty}^{\infty} F(\xi)^*F(\xi)\mathrm{d}\xi \equiv \int_{-\infty}^{\infty} |F(\xi)|^2 \mathrm{d}\xi$$

Therefore:

$$\int_{-\infty}^{\infty} |f(x)|^2 dx = \int_{-\infty}^{\infty} |F(\xi)|^2 d\xi$$

If $f(x)$ represents the complex amplitude of a wave, or a static distribution of illumination, then $|f(x)|^2$ corresponds to the energy density (in this 1D case, the energy per unit length), and its integral gives the total energy. Consequently, Parseval's theorem implies that $\int_{-\infty}^{\infty} |F(\xi)|^2 d\xi$ is also the total energy, where $|F(\xi)|^2$ represent the energy density in the frequency domain. For this reason, $|F(\xi)|^2$ is often referred to as the power spectrum.

A simple case that allows to verify Parseval's theorem is the Gaussian function $f(x) = \exp(-\pi x^2)$. For this function:

$$\int_{-\infty}^{\infty} |f(x)|^2 dx = \int_{-\infty}^{\infty} \exp(-2\pi x^2) dx$$

Let $u = \sqrt{2\pi}\, x$, so $du = \sqrt{2\pi}\, dx$. The integral becomes:

$$\int_{-\infty}^{\infty} \exp(-2\pi x^2) dx = \frac{1}{\sqrt{2\pi}} \int_{-\infty}^{\infty} \exp(-u^2) du$$

Recalling from section 3.5.2 that $\int_{-\infty}^{\infty} \exp(-u^2) du = \sqrt{\pi}$.

$$\int_{-\infty}^{\infty} |f(x)|^2 dx = \frac{1}{\sqrt{2}}$$

As calculated in the section 3.5.2, the Fourier transform of $f(x) = \exp(-\pi x^2)$ is the Gaussian function $F(\xi) = \exp(-\pi \xi^2)$, it follows that:

$$\int_{-\infty}^{\infty} |F(\xi)|^2 d\xi = \frac{1}{\sqrt{2}}$$

Then, $\int_{-\infty}^{\infty} |f(x)|^2 dx = \int_{-\infty}^{\infty} |F(\xi)|^2 d\xi = \frac{1}{\sqrt{2}}$.

7.3.2 The Parseval's theorem and the autocorrelation

Parseval's theorem provides an interesting perspective on autocorrelation. If we consider a complex function, the autocorrelation $C_{ff}(X)$ is given by:

$$C_{ff}(X) = \int_{-\infty}^{\infty} f(x) f(x - X)^* dx$$

Where $f(x - X)^*$ denotes the complex conjugate of $f(x - X)$. To simplify calculations, we can apply a change of variable. Let $u = x - X$ which implies $du = dx$. Substituting this into the integral, the autocorrelation can be written as:

$$C_{ff}(X) = \int_{-\infty}^{\infty} f(u + X) f(u)^* du$$

Expressing $f(u)^*$ in terms of the Fourier transform, we have:

$$f(u)^* = \int_{-\infty}^{\infty} F^*(\xi)\exp(-i2\pi\xi u)d\xi$$

Substituting this into the autocorrelation integral, we obtain:

$$C_{ff}(X) = \int_{-\infty}^{\infty} f(u+X)\left[\int_{-\infty}^{\infty} F^*(\xi)\exp(-i2\pi\xi u)d\xi\right]du$$

Rearranging the order of integration, this becomes:

$$C_{ff}(X) = \int_{-\infty}^{\infty} F^*(\xi)\left[\int_{-\infty}^{\infty} f(u+X)\exp(-i2\pi\xi u)du\right]d\xi$$

The inner integral is the Fourier transform of $f(u+X)$. Using the translation theorem (equation (3.14)), the Fourier transform of $f(u+X)$ is:

$$F\{f(u+X)\} = F(\xi)\exp(i2\pi\xi X)$$

Substituting this result back into the expression for $C_{ff}(X)$, we obtain:

$$C_{ff}(X) = \int_{-\infty}^{\infty} F^*(\xi)F(\xi)\exp(i2\pi\xi X)d\xi = \int_{-\infty}^{\infty} |F(\xi)|^2 \exp(i2\pi\xi X)d\xi$$

Here, the integral $\int_{-\infty}^{\infty} |F(\xi)|^2 \exp(+i2\pi\xi X)d\xi$ is the inverse Fourier transform of $|F(\xi)|^2$. Therefore, we can write:

$$C_{ff}(X) = F^{-1}\{|F(\xi)|^2\}$$

Taking the Fourier transform to both terms:

$$F\{C_{FF}(X)\} = |F(\xi)|^2$$

This relationship is known as the Wiener–Khinchin theorem. In simple terms, it states that the Fourier transform of the autocorrelation of a function is equal to its power spectrum.

As an example of the application of the Wiener–Khinchin theorem consider the autocorrelation of two rectangular functions as determined in section 7.1. This autocorrelation resulted in a triangular function. Furthermore, the Fourier transform of the triangular function yields a sinc² (see problem 3.3) which represents the power spectrum of the rectangular function.

7.4 The optical transfer function

As discussed in chapter 5 (equation (5.7)), the image produced by an optical system can be described as the convolution of the object function $f(x)$ (or $f(x, y)$ in 2D) and the PSF $h(x)$ (or $h(x, y)$ in 2D):

$$g(X) = (f \otimes h)(X) \tag{7.6}$$

Here, $g(X)$ represents the resulting image. For a perfect, ideal optical system, the PSF $h(x) = \delta(x)$, and since the convolution with a delta function results in the function itself:

$$g = f$$

In this case, the image would be an exact replica of the object. To further analyse this, we take the Fourier transform of both sides of equation (7.6):

$$F\{g(X)\} = F\{f \otimes h(X)\}$$

Applying the convolution theorem (equation (5.12)), we obtain:

$$F\{g(X)\} = F\{f(x)\}F\{h(x)\}$$

For a perfect image $h(x) = \delta(x)$, it follows that $F\{h(x)\} = 1$, resulting in:

$$F\{g(X)\} = F\{f(x)\}$$

as expected. This confirms that the image in the frequency domain is identical to the object in the frequency domain for an ideal system.

The term $F\{h(x)\}$ is known as the Optical Transfer Function (OTF(ξ)). It represents the effect of the optical system on the frequency spectrum of the image. In the spatial domain $h(x)$ describe the system's response, while in the frequency domain, the OTF(ξ) serves the equivalent role. OTF(ξ) is, in general, a complex function that can be expressed as:

$$\text{OTF}(\xi) = \text{MTF}(\xi)\exp\left(i\text{PTF}(\xi)\right) \tag{7.7}$$

Here, MTF(ξ) refers to the Modulation Transfer Function, and PTF(ξ) refers to the Phase Transfer Function. Together, these components form the optical transfer function OTF(ξ), which fully characterizes the influence of the optical system on both the amplitude and phase of the object spatial frequencies.

7.4.1 Effect of a linear optical system on a specific frequency component

To understand how a linear optical system affects a specific frequency component, consider the function $f(x)$ that represents the intensity of a harmonic component of the object. The function has amplitude A, frequency ξ_0 and phase ϕ:

$$f(x) = 1 + A\cos(2\pi\xi_0 x + \phi) \tag{7.8}$$

When this function passes through an optical system, the resulting image $g(X)$ is the convolution of the $f(x)$ component with the system spread function (in this 1D case, the line spread function $h(x)$):

$$g(X) = [1 + A\cos(2\pi\xi_0 x + \phi)] \otimes h(x)$$

Using the convolution theorem (equation (5.12)) the Fourier transform of $g(X)$ is given by:

$$F\{g(X)\} = F\{1 + A\cos(2\pi\xi_0 x + \phi)\} \cdot F\{h(x)\}$$

Since $F\{h(x)\}$ was defined as OTF(ξ), and using the Euler formula, we can write:

$$F\{g(X)\} = \left[\delta(\xi) + F\left\{\frac{1}{2}A(\exp(i(2\pi\xi_0 x + \phi)) + \exp(-i(2\pi\xi_0 x + \phi)))\right\}\right] \cdot \text{OTF}(\xi)$$

Taking the Fourier transform of the exponential terms:

$$F\{g(X)\} = \left[\delta(\xi) + \frac{1}{2}A(\exp(i\phi)\delta(\xi - \xi_0) + \exp(-i\phi)\delta(\xi + \xi_0))\right] \cdot \text{OTF}(\xi)$$

Expanding this:

$$F\{g(X)\} = \text{OTF}(0)\delta(\xi) + \frac{1}{2}A(\exp(i\phi)\text{OTF}(\xi_0)\delta(\xi - \xi_0) + \exp(-i\phi)\text{OTF}(-\xi_0)\delta(\xi + \xi_0))$$

To find $g(x)$, we take the inverse Fourier transform:

$$g(X) = F^{-1}\left\{\text{OTF}(0)\delta(\xi) + \frac{1}{2}A(\exp(i\phi)\,\text{OTF}(\xi_0)\delta(\xi - \xi_0) + \exp(-i\phi)\,\text{OTF}(-\xi_0)\delta(\xi + \xi_0))\right\}$$

$$g(X) = \int_{-\infty}^{\infty} \text{OTF}(0)\delta(\xi)\exp(i2\pi\xi X)d\xi$$
$$+ \int_{-\infty}^{\infty} \frac{A}{2}\exp(i\phi)\,\text{OTF}(\xi_0)\delta(\xi - \xi_0)\exp(i2\pi\xi X)d\xi$$
$$+ \int_{-\infty}^{\infty} \frac{A}{2}\exp(-i\phi)\,\text{OTF}(-\xi_0)\delta(\xi + \xi_0)\exp(i2\pi\xi X)d\xi$$

Since $\text{OTF}(-\xi_0) = \text{OTF}^*(\xi_0)$:[1]

$$g(X) = \text{OTF}(0) + \frac{A}{2}\exp(i\phi)\,\text{OTF}(\xi_0)\exp(i2\pi\xi_0 X) + \frac{A}{2}\exp(-i\phi)\,\text{OTF}^*(\xi_0)\exp(-i2\pi\xi_0 X)$$

Here, substituting $\text{OTF}(\xi_0) = \text{MTF}(\xi_0)\exp(i\text{PFT}(\xi_0))$ and $\text{OTF}^*(\xi_0) = \text{MTF}^*(\xi_0)\exp(-i\text{PFT}(\xi_0))$ as defined in equation (7.7):

$$g(X) = \text{OTF}(0) + \frac{A}{2}\exp(i\phi)\,\text{MTF}(\xi_0)\exp(i\text{PFT}(\xi_0))\exp(i2\pi\xi_0 X)$$
$$+ \frac{A}{2}\exp(-i\phi)\,\text{MTF}(\xi_0)\exp(-i\text{PFT}(\xi_0))\exp(-i2\pi\xi_0 X)$$

Here, we have used $\text{MTF}^*(\xi_0) = \text{MTF}(\xi_0)$ since MTF is real. Thus:

$$g(X) = \text{OTF}(0) + \frac{A}{2}\text{MTF}(\xi_0)[\exp(i2\pi\xi_0 X + i\phi + i\text{PFT}(\xi_0))$$
$$+ \exp(-(i2\pi\xi_0 X + i\phi + i\text{PFT}(\xi_0)))]$$

Using the Euler formula for the cosine function:

$$g(X) = \text{OTF}(0) + A\,\text{MTF}(\xi_0)\cos[2\pi\xi_0 X + \phi + \text{PFT}(\xi_0)] \qquad (7.9)$$

Here, OTF(0) is the value of the Fourier transform of the PSF at $\xi = 0$ which is a real constant given by $\text{OTF}(0) = \int_{-\infty}^{\infty} h(x)dx$.

---

[1] $\text{OTF}^*(\xi_0) = \left[\int_{-\infty}^{\infty} h(x)\exp(-i2\pi\xi_0 x)dx\right]^*$; as $h(x)$ is real $\text{OTF}^*(\xi_0) = \int_{-\infty}^{\infty} h(x)\exp(i2\pi\xi_0 x)dx$
$= \int_{-\infty}^{\infty} h(x)\exp(-i2\pi\xi_0(-x))dx = \text{OTF}(-\xi_0)$.

By comparing equations (7.8) and (7.9), it can be concluded that the effect of the optical system over any harmonic component is threefold: (i) it multiplies the amplitude by the modulus of the OTF at ξ_0 (MTF(ξ_0)), (ii) it adds the phase of the OTF at ξ_0 (PTF(ξ_0)) to the phase of the harmonic, and (iii) it alters the zero frequency term (often called DC term, for Direct Current, borrowed from electronics). Importantly, the image function retains the harmonic character and frequency of the object's component.

7.4.2 Modulation transfer function and visibility

In figure 7.4 a hypothetical example illustrates this effect: the optical system increases the phase by PTF(ξ_0), scales the amplitude by MTF(ξ_0) = 0.5, and decreases the DC term into OTF(ξ_0) = 0.3 units. As observed, the frequency component becomes less visible in the image compared to the object.

The visibility, defined as:

$$V = \frac{I_{\max} - I_{\min}}{I_{\max} + I_{\min}}$$

is a key measure of contrast. According to this definition, the visibility of the pattern of this harmonic is 1 for the object and 0.71 for the image. For instance, if this harmonic pattern is the sole frequency in the object (e.g. a sinusoidal grid), the image will not contain any areas that are completely dark, and its maximum intensity will be lower than that of the object.

This reduction in visibility results from the smoothing effect of the spread function, whose convolution with the object blurs the image. The smoothing effect becomes more pronounced at higher frequencies, where the size of the spread function is comparable to the spatial period of the harmonic.

Consequently, the MTF, and hence the visibility, decreases with increasing frequency. Figure 7.5 illustrates a typical MTF curve as a function of frequency, which serves as a critical figure of merit for evaluating optical system performance.

To conclude this section, a few words about the PTF. The PTF is generally less significant than the MTF and, in many cases, may be disregarded. It arises because

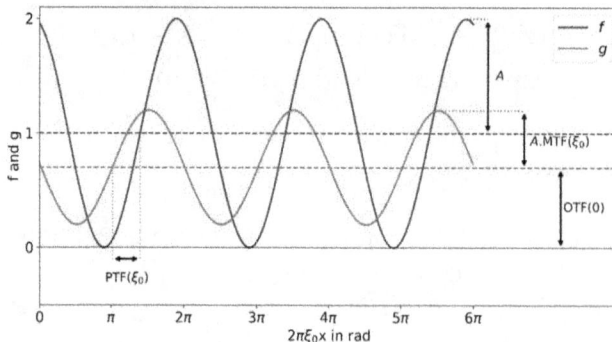

Figure 7.4. The intensity as a function of the phase for a harmonic in the object (blue) and in the image (red).

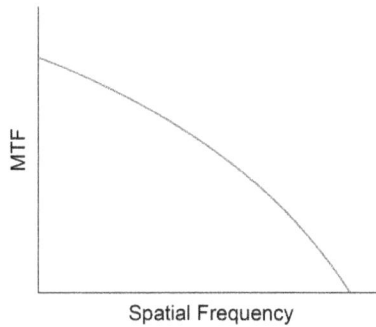

Figure 7.5. A typical curve illustrating the dependence of the MTF on frequency for a hypothetical optical system.

of a non-symmetric (non-even) spread function, typically caused by optical aberrations or misalignment within the optical system. Obviously, a symmetric spread function cannot introduce a phase change in the image. This is because the Fourier transform of a real and even function is itself real. Therefore, the OTF corresponding to an even spread function is real, and its PTF is zero.

7.5 The image formation: the lens and the Fourier transform. The 4F system

As previously discussed, the distribution of the electric field on a screen placed at infinity (Fraunhofer diffraction) corresponds to the Fourier transform of the aperture function. A beam of parallel rays, which would naturally converge at infinity, can instead be focussed at a finite distance by introducing a lens in front of the diffracting object and placing the screen at the rear focal plane of the lens (see section 2.2). On this screen, the diffraction pattern emerges. In this configuration, it can be stated that the lens effectively performs the Fourier transform of the aperture function. This concept is visually represented in figure 7.6.

Figure 7.7 shows a sketch of the so-called 4F system. In this setup, the diffracted beams originating in the object plane reach a lens positioned at its focal distance from the object. This lens focusses the beams and produces a diffraction pattern (so, a Fourier transform) at the Fourier transform (*FT*) plane, which is located at its rear focal plane. However, there is no screen at this plane, so the beams, after converging, continue their path towards a second lens, placed at the focal distance of the *FT* plane. This second lens performs the Fourier transform, projecting the result onto a screen located at its rear focal plane. As a whole, the system performs two consecutive Fourier transforms, which, as calculated in section 3.3.4, results in an inverted image of the object.

Not all spatial frequency beams can be collected by the first lens. High spatial frequencies, diffracted at large angles relative to the optical axis, may be lost, leading to a reduction in image contrast. This is because high spatial frequencies are critical for reproducing sharp, abrupt variation in the object's illumination. Consequently,

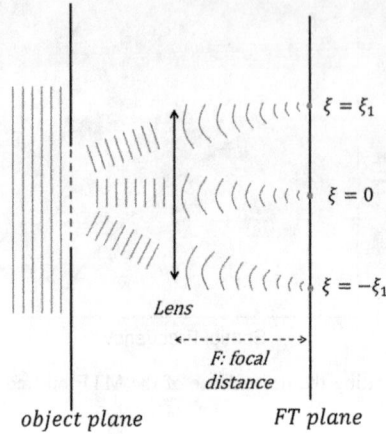

Figure 7.6. The lens focusses parallel beams, corresponding to different spatial frequencies, onto its focal plane. In this figure, three beams are shown, represented by their wavefronts, being diffracted by the object in three different directions. As previously discussed, each diffracted direction corresponds to a specific spatial frequency (ξ).

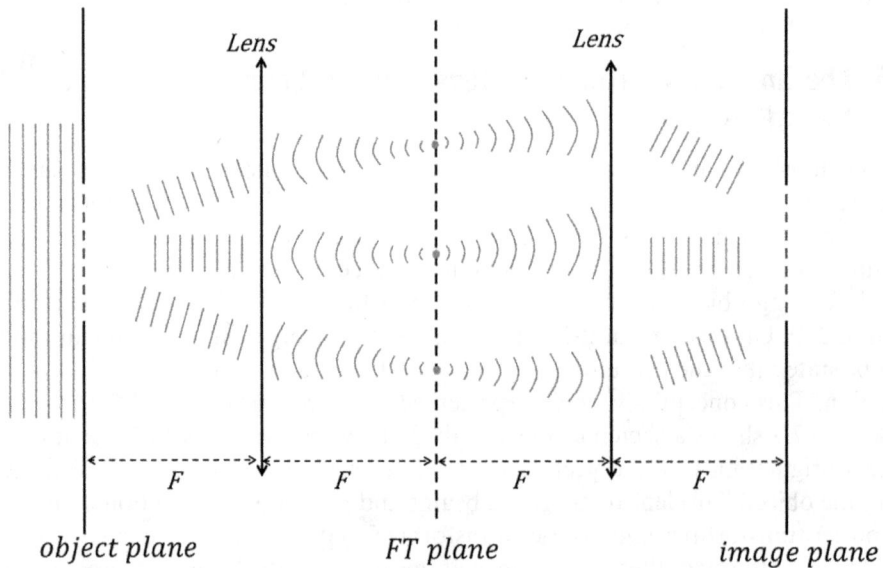

Figure 7.7. The 4F system.

the 4F system acts as a low-pass filter. Hence, the lens diameter significantly impacts image quality: a larger lens improves image contrast.

7.5.1 The 4F system as a filtering device

The 4F system can be used for filtering purposes. By placing appropriate masks in the *FT* plane specific frequencies can be eliminated.

7.5.2 The 4F system as correlator

Another application of the 4F system is as correlator. In this application, a transparency containing the input function $f(x, y)$ is placed in the object plane. The pattern passes through the first lens and its Fourier transform $F(\xi, \eta)$ is obtained in the *FT* plane. At this plane, a filter is placed which contains the pattern of the Fourier transform $H(\xi, \eta)$ of the function $h(x, y)$ which needs to be compared with $f(x, y)$.

In this process, the 4F system performs the product of both Fourier transforms. To understand this, consider two different transparencies, one placed after the other. If the first transparency transmits, say, 0.6 of the incoming intensity, and the second transmits the 0.5, the result is 0.3, which corresponds to 0.6 times 0.5. Then, from the *FT* planes, emerges towards the second lens, the product of the Fourier transform of both images.

According to the convolution theorem (see equation (5.12)), this product represents the Fourier transform of the convolution of $f(x, y)$ by $h(x, y)$:

$$F\{f(x, y)\}F\{h(x, y)\} = F\{(f \otimes h)(x, y)\}$$

After passing through the second lens, we have the Fourier transform of that previous quantity. We have seen in section 3.3.4 that the Fourier transform of the Fourier Transform of a function is the function itself with a minus sign in the argument. Then:

$$F\{F\{f(x, y)\}F\{h(x, y)\}\} = F\{F\{(f \otimes h)(x, y)\}\} = (f \otimes h)(-x, -y),$$

In the end, this Fourier transform of the product of the Fourier transforms yields the convolution of $f(x, y)$ and $h(x, y)$.

However, if the filter placed in the *FT* plane is the conjugate $H^*(\xi, \eta)$ instead of $H(\xi, \eta)$, the operation produces the correlation rather than the convolution after the second lens (see problem 7.1). In this case, the 4F systems acts as a correlator of the functions $f(x, y)$ and $h(x, y)$.

7.6 The evolution of the complex amplitude

7.6.1 The angular spectrum of plane waves

For a monochromatic plane wave propagating in an arbitrary direction, the complex amplitude pattern in a plane normal to the z-axis (at $z = 0$) is given by (see section 2.4.6):

$$\mathcal{E}(x, y, z = 0) = A \exp\left(i(2\pi\xi x + 2\pi\eta y)\right)$$

Here, $2\pi\xi$ and $2\pi\eta$ represent the components k_x and k_y of the wavevector $\vec{k}$, whose magnitude is $k = \frac{2\pi}{\lambda}$. Thus, k_x and k_y, besides being the angular spatial frequency of the complex amplitude in the given plane, also allow us to define a plane wave:

$$\mathcal{E}(x, y, z) = A \exp\left(i(2\pi\xi x + 2\pi\eta y + k_z z)\right)$$

This is because, once k_x and k_y are known, k_z can be determined using:

$$k_x^2 + k_y^2 + k_z^2 = k^2 \equiv \left(\frac{2\pi}{\lambda}\right)^2$$

Thus, k_z can be obtained as:

$$k_z = \pm\sqrt{\left(\frac{2\pi}{\lambda}\right)^2 - k_x^2 - k_y^2}$$

The only indeterminacy in k_z is the sign. We will assume that k_z is oriented toward the positive direction of the z-axis, and thus, the sign is positive (choosing k_z in the opposite direction will not change the general result). Therefore, the complex amplitude of the plane wave defined by k_x and k_y, can be written as:

$$\underline{\mathcal{E}}(x, y, z) = A \exp\left(i\left[2\pi\xi x + 2\pi\eta y + \left(\sqrt{\left(\frac{2\pi}{\lambda}\right)^2 - k_x^2 - k_y^2}\right)z\right]\right) \qquad (7.10)$$

We can say that each pair (k_x, k_y) defines a plane wave in the direction of the vector $\vec{k}$. This direction is determined by the direction cosines of the vector $\vec{k}$:

$$\cos\alpha = \frac{k_x}{k}, \quad \cos\beta = \frac{k_y}{k}, \quad \text{and } \cos\gamma = \frac{k_z}{k}$$

In the same way as k_z is determined by knowing k_x and k_y (up to the sign), also $\cos\gamma$ can be determined since:

$$\cos^2\alpha + \cos^2\beta + \cos^2\gamma = \left(\frac{k_x}{k}\right)^2 + \left(\frac{k_y}{k}\right)^2 + \left(\frac{k_z}{k}\right)^2 = 1$$

And this equation can be solved for $\cos\gamma$ (up to the sign).

The previous considerations enable an insightful interpretation of any general wave, which is typically not a plane wave. The Fourier transform of such a wave $\underline{\mathcal{E}}(x, y, z)$, in the plane $z = 0$ is:

$$F\{\underline{\mathcal{E}}(x, y, 0)\} \equiv F(\xi, \eta) = \iint_{-\infty}^{\infty} \underline{\mathcal{E}}(x, y, 0) \exp(-i(2\pi\xi x + 2\pi\eta y)) dx dy$$

Taking the inverse Fourier transform leads to:

$$F^{-1}\{F(\xi, \eta)\} \equiv \underline{\mathcal{E}}(x, y, 0) = \iint_{-\infty}^{\infty} F(\xi, \eta) \exp(i(2\pi\xi x + 2\pi\eta y)) d\xi d\eta \qquad (7.11)$$

In this equation, the functions $F(\xi, \eta) \exp(i(2\pi\xi x + 2\pi\eta y)) d\xi d\eta$ represent 2D harmonics confined to the $z = 0$ plane. However, according to the previous analysis, these harmonics exhibit a one-to-one correspondence with plane waves propagating along different directions. Thus, $F(\xi, \eta) d\xi d\eta$ simultaneously describes both the amplitude of these 2D harmonics and the amplitude of their associated plane waves, that can be expressed as:

$$F(\xi, \eta) \exp\left(i\left[2\pi\xi x + 2\pi\eta y + \left(\sqrt{\left(\frac{2\pi}{\lambda}\right)^2 - k_x^2 - k_y^2}\right)z\right]\right)d\xi d\eta \qquad (7.12)$$

For this reason, equation (7.11) can be interpreted as follows: the complex amplitude in a plane is formed by the sum (integral) of plane waves oriented in different directions. In other words, any wave can be represented as a superposition of plane waves propagating in different directions. Since k_x and k_y define the direction of the plane waves, $F(\xi, \eta)$ is interpreted as the angular spectrum of plane waves.

This interpretation is interesting because it allows one to predict the evolution of the complex amplitude $\mathcal{E}(x, y, z)$ through an optical system, by analysing the evolution of the component plane waves and then reconstructing the evolved wave by means of the inverse Fourier transform.

7.6.2 The Transfer function of free space

As demonstrated in section 7.4.1, the effect of a linear and spatially invariant optical system over a harmonic component is to multiply the amplitude of the harmonic by the magnitude of the OTF (namely MTF(ξ, η)), and add the phase of the OTF (namely PTF(ξ, η)) to the phase of the harmonic[2]. Then, the harmonic:

$$\exp\left(i(2\pi\xi x + 2\pi\eta y)\right) \qquad (7.13)$$

is transformed by the optical system in:

$$\mathrm{MTF}(\xi, \eta)\exp\left(i(2\pi\xi x + 2\pi\eta y)\right)\exp\left(i\mathrm{PTF}(\xi, \eta)\right) \qquad (7.14)$$

In free space we can consider that the harmonic input of the optical system in the object plane is:

$$\exp\left(i(2\pi\xi x + 2\pi\eta y)\right) \qquad (7.15)$$

And the harmonic output at a parallel plane at $z = z_1$ is:

$$\exp\left(i\left[2\pi\xi x + 2\pi\eta y + \left(\sqrt{\left(\frac{2\pi}{\lambda}\right)^2 - k_x^2 - k_y^2}\right)z_1\right]\right) \qquad (7.16)$$

Comparing equations (7.16) and (7.14), it can be concluded that the transfer function in this case is:

$$OTF(\xi, \eta) = \exp\left(i\left(\sqrt{\left(\frac{2\pi}{\lambda}\right)^2 - k_x^2 - k_y^2}\right)z_1\right)$$

That is, the amplitude of the harmonics is not modified, only the phase.

---

[2] In section 7.4.1, the used harmonic function was $1 + A\cos(2\pi\xi_0 x + \phi)$, but the same result can be obtained by using the complex representation in two dimensions: $\mathcal{E}(x, y) = A \exp\left(i(2\pi\xi x + 2\pi\eta y)\right)$.

7.6.3 The evanescent waves

It can be observed that the extent of k_x and k_y is, in principle, not limited. It depends on the distribution of the complex amplitude in the given z-plane. For example, if there is a narrow aperture in the z-plane, the k_x and k_y spectra can become very wide and k_x and k_y can reach very large values.

For these high values of k_x and k_y, the radicand of equation (7.12) can become negative if $(\frac{2\pi}{\lambda})^2$ is smaller than $k_x^2 + k_y^2$. In this case, the square root becomes imaginary:

$$\exp\left(i\left[2\pi\xi x + 2\pi\eta y + \left(i\sqrt{k_x^2 + k_y^2 - \left(\frac{2\pi}{\lambda}\right)^2}\right)z\right]\right)$$

This can simplify to:

$$\exp\left(i(2\pi\xi x + 2\pi\eta y)\right)\exp\left(-\sqrt{k_x^2 + k_y^2 - \left(\frac{2\pi}{\lambda}\right)^2}\,z\right)$$

In this equation, the factor $\exp(-\sqrt{k_x^2 + k_y^2 - (\frac{2\pi}{\lambda})^2}\,z)$ does not represent an oscillating field but a decaying field amplitude wave.

This is called an evanescent wave. These high spatial frequency components do not propagate over large distances but decay very close to the $z = 0$ plane. High spatial frequencies are necessary to reproduce very small details, and for this reason, their existence represents a limit (the diffraction limit) for resolution. However, today there are techniques for detecting these evanescent waves before they disappear, allowing us to obtain information about tiny details. This is called super-resolution and is used in near-field microscopy.

7.7 Problems

Problem 7.1 Demonstrate the theorem of the correlation:

$$F\{(f \circledast h)(X)\} = F\{f(x)\} \cdot F\{h(x)\}^*$$

Problem 7.2 (with solution)
Verify the Parseval's theorem for the function $f(x) = \exp(-|x|)$
Solution:
The Parseval's theorem states: $\int_{-\infty}^{\infty} |f(x)|^2 dx = \int_{-\infty}^{\infty} |\hat{f}(\xi)|^2 d\xi$
Calculation of the Fourier transform of $f(x) = \exp(-|x|)$

$$F(\xi) = \int_{-\infty}^{\infty} \exp(-|x|)\exp(-i2\pi\xi x)dx$$

To solve this integral it must be divided into two parts:

$$F(\xi) = \int_{-\infty}^{0} \exp(x)\exp(-i2\pi\xi x)dx + \int_{0}^{\infty} \exp(-x)\exp(-i2\pi\xi x)dx$$

This is because $-|x| = x$ if $x \leqslant 0$ and $-|x| = -x$ if $x \geqslant 0$

Solving the integrals:

$$F(\xi) = \left. \frac{\exp(x - i2\pi\xi x)}{1 - i2\pi\xi} \right|_{-\infty}^{0} - \left. \frac{\exp(-x - i2\pi\xi x)}{1 + i2\pi\xi} \right|_{0}^{\infty}$$

Evaluating in zero, both numerators become 1. Evaluation in $x = -\infty$ in the first term and $x = \infty$ in the second term gives 0. Thus:

$$F(\xi) = \frac{1}{1 - i2\pi\xi} + \frac{1}{1 + i2\pi\xi}$$

Now, let us calculate

$$\int_{-\infty}^{\infty} |F(\xi)|^2 d\xi = \int_{-\infty}^{\infty} \left[\frac{1 + i2\pi\xi + 1 - i2\pi\xi}{(1 - i2\pi\xi)(1 + i2\pi\xi)} \right]^2 d\xi$$

Simplifying the expression:

$$\int_{-\infty}^{\infty} |F(\xi)|^2 d\xi = \int_{-\infty}^{\infty} \left(\frac{2}{1 + (2\pi\xi)^2} \right)^2 d\xi = \int_{-\infty}^{\infty} \frac{4}{(1 + (2\pi\xi)^2)^2} d\xi$$

Let $u = 2\pi\xi$, so $d\xi = \frac{du}{2\pi}$. Then:

$$\int_{-\infty}^{\infty} |F(\xi)|^2 d\xi = \int_{-\infty}^{\infty} \frac{4}{(1 + u^2)^2} \cdot \frac{du}{2\pi} = \frac{4}{2\pi} \int_{-\infty}^{\infty} \frac{1}{(1 + u^2)^2} du$$

This is a known integral and its value is $\frac{\pi}{2}$. Then:

$$\int_{-\infty}^{\infty} |F(\xi)|^2 d\xi = 1$$

Now, let us calculate $\int_{-\infty}^{\infty} |f(x)|^2 dx$:

$$\int_{-\infty}^{\infty} |f(x)|^2 dx = \int_{-\infty}^{\infty} \exp(-2 | x |) dx$$

We now divide the integral in two parts:

$$\int_{-\infty}^{0} \exp(2x) dx + \int_{0}^{\infty} \exp(-2x) dx$$

Solving the integrals:

$$\left. \frac{\exp(2x)}{2} \right|_{-\infty}^{0} - \left. \frac{\exp(-2x)}{2} \right|_{0}^{\infty} = \frac{1}{2} + \frac{1}{2} = 1$$

Thus: $\int_{-\infty}^{\infty} |f(x)|^2 dx = \int_{-\infty}^{\infty} |F(\xi)|^2 d\xi$.

Further Reading

[1] Goodman J W 2017 *Introduction to Fourier Optics* 4th edn (New York: W.H. Freeman)

[2] Born M and Wolf E 1999 *Principles of Optics* 7th edn (Cambridge: Cambridge University Press)

[3] Mahajan V N 2019 *Optical Imaging and Aberrations: Part I. Ray Geometrical Optics* 2nd edn (Bellingham, WA: SPIE Press)

[4] Khare K 2015 *Fourier Optics and Computational Imaging* (Hoboken, NJ: Wiley)

[5] Saleh B E A and Teich M C 2019 *Fundamentals of Photonics* 3rd edn (Hoboken, NJ: Wiley)

Chapter 8

Advanced topics

This chapter presents advanced techniques and applications within the field of this book, focussing on three key topics: Fourier transform infrared spectroscopy (FTIR), numerical algorithms for computing Fourier transforms, and holography. FTIR is explored as an important tool in spectral analysis, highlighting its advantages over other spectroscopy techniques. The chapter then moves to numerical approaches used in Fourier optics, addressing first the problem of sampling and the Whittaker–Shannon theorem, then, the discrete Fourier transform (DFT) and finally the fast Fourier transform (FFT) algorithm. Finally, we present the remarkable concept of holography. It consists in recording not only the intensity of light, as in a photograph, but also its phase through interference patterns. This enables the reconstruction of the full electromagnetic field, creating a three-dimensional effect. We first outline the general principle, then rigorously derive the equations.

8.1 Fourier transform infrared spectroscopy

8.1.1 Optical spectroscopies

Optical spectroscopy is a versatile and essential tool in science and technology. It enables the study of light–matter interactions and provides valuable insights into the structure, optical properties, and electronic characteristics of the materials. Through spectroscopy, it is possible to identify crystal or molecular structures in both terrestrial and celestial objects, develop quality control systems across diverse industries, and aid in medical diagnostics for a range of human conditions. Spectroscopy is applied across the entire electromagnetic spectrum (figure 8.1), spanning from radio waves to gamma rays.

Spectra represent the dependence of a measured quantity—typically intensity or counts, depending on the type of detector—on the energy of the photons or the wavelength of the measured radiation. In addition to photons, other quantum particles, such as electrons, can also be utilized in spectroscopy. In this section, we will focus on optical spectroscopy, specifically infrared (IR) spectroscopy.

Frequency (hz)

10^2 10^4 10^6 10^8 10^{10} 10^{12} 10^{14} 10^{16} 10^{18} 10^{20} 10^{22}

Radio   infrared ultraviolet gamma

microwaves   x-rays

visible

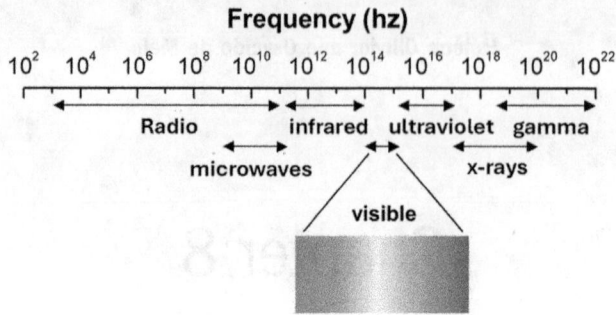

Figure 8.1. The electromagnetic spectrum.

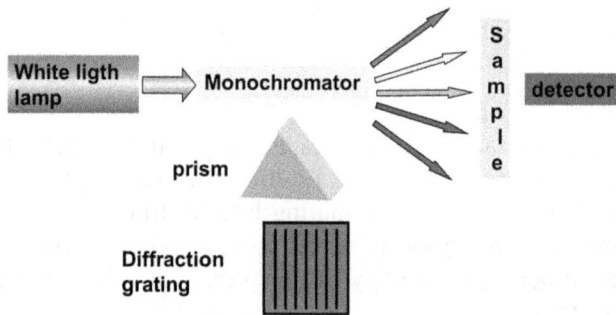

White ligth lamp ⇒ Monochromator

prism

Diffraction grating

Sample

detector

Figure 8.2. Simplified sketch of a traditional setup for measuring optical transmission spectra.

IR spectroscopy explores the range in which molecular vibrational spectra or crystal phonons occur, enabling the identification of specific molecules or phases. The IR range is divided into near-IR (800–2500 nm wavelength range), mid-IR (2.5–15 μm), and far-IR (15–250 μm). It is also common to use wavenumber ($\frac{1}{\lambda}$, spatial frequency) instead of wavelength.

A simplified view of the setup commonly used for traditional optical spectroscopy is shown in figure 8.2. The white light lamp emits a broad spectrum, which is selected according to the required measurement range. For the visible spectrum, a tungsten lamp is typically used, while for the IR range, a black body radiator can be employed. A monochromator disperses the light beam into its different wavelength components using a prism or a diffraction grating. Between the monochromator and the sample, a narrow slit (not shown) allows only a small range of wavelengths $\Delta\lambda$ to pass through to the sample. For transmission measurements, a detector is placed behind the sample to measure the transmitted intensity. By rotating the prism or diffraction grating, it is possible to scan the entire wavelength range and obtain the transmission spectrum. Different types of detectors are used for different wavelength ranges, and various geometries can be employed to measure other types of spectra, such as reflection, Raman or photoluminescence spectra.

8.1.2 Fourier transform infrared spectroscopy (FTIR)

FTIR is an excellent application of the concepts studied in this book. A simplified view of the main components of an FTIR spectrometer is sketched in figure 8.3. This technique uses a Michelson interferometer to obtain optical spectra in the infrared region. Unlike traditional spectroscopy, FTIR does not rely on prisms or diffraction gratings to disperse wavelengths in the spectra. The light beam whose spectrum is to be measured (e.g., the spectrum of a lamp or the transmitted or reflected spectrum from a sample) is directed toward a beam splitter, which divides the beam into two beams of equal intensity. One beam is reflected toward a movable mirror, while the other is transmitted toward a fixed mirror. After reflecting off the mirrors, both beams return to the beam splitter, where one is transmitted, and the other is reflected toward the detector.

The phase difference between two beams with equal spatial frequency ξ (spatial period or wavelength λ) that interfere (see the conditions for interference in section 2.4.3) is given by:

$$\Delta\Phi = 2\pi\xi x,$$

where $x = x_2 - x_1$ is the path difference between the beams. This path difference is varied by displacing the movable mirror. When the beams arrive at the detector, their electric fields overlap, and the resultant complex amplitude is:

$$\underline{\mathcal{E}}(x_1, x_2, \xi) = \underline{\mathcal{E}}_1(x_1, \xi) + \underline{\mathcal{E}}_2(x_2, \xi) \tag{8.1}$$

As the initial beam is split into two beams of equal intensity by the beam splitter, each beam carries half of the original intensity ($I/2$) and, since the intensity I is proportional to the square of the electric field amplitude, the amplitude of each divided beam is reduced by a factor of $\sqrt{2}$. Thus:

$$\underline{\mathcal{E}}_1(x_1, \xi) = \frac{\underline{\mathcal{E}}_0(\xi)}{\sqrt{2}} \exp(i2\pi\xi x_1) \text{ and } \underline{\mathcal{E}}_2(x_2, \xi) = \frac{\underline{\mathcal{E}}_0(\xi)}{\sqrt{2}} \exp(i2\pi\xi x_2)$$

where we factorized out the phase term $\exp(i2\pi\xi x_1)$ which we omitted since it cancels out in the intensity calculation anyway. $\underline{\mathcal{E}}_0(\xi)$ is the incident electric field. Thus, equation (8.1) becomes:

$$\underline{\mathcal{E}}(x, \xi) = \frac{\underline{\mathcal{E}}_0(\xi)}{\sqrt{2}}(1 + \exp(i2\pi\xi x)) \tag{8.2}$$

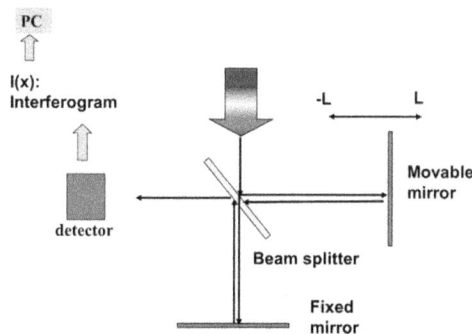

Figure 8.3. Basic elements of an FTIR spectrometer.

The intensity at the detector is given by expression (2.8):

$$I(x, \xi) = \frac{K}{2}\underline{\mathcal{E}}(x, \xi)\underline{\mathcal{E}}^*(x, \xi)$$

where $I(x, \xi)$ is called the spectral intensity and corresponds with the intensity of an infinitesimal spatial frequency range between ξ and $\xi + d\xi$, divided by $d\xi$.

Injecting the expression of the complex amplitude (equation (8.2)):

$$I(x, \xi) = \frac{K}{2}\left[\frac{\underline{\mathcal{E}}_0(\xi)}{\sqrt{2}}(1 + \exp(i2\pi\xi x))\right]\left[\frac{\underline{\mathcal{E}}_0^*(\xi)}{\sqrt{2}}(1 + \exp(-i2\pi\xi x))\right]$$

And using the Euler formula:

$$I(x, \xi) = \frac{K}{2}\frac{|\underline{\mathcal{E}}_0(\xi)|^2}{2}2[1 + \cos(2\pi\xi x)] = I_0(\xi)[1 + \cos(2\pi\xi x)]$$

where:

$$I_0(\xi) = \frac{K}{2}|\underline{\mathcal{E}}_0(\xi)|^2$$

represents the spectral intensity of the initial beam, i.e., the spectrum we want to measure. Around this value, $I(x, \xi)$ oscillates with x.

Let us now calculate the total intensity measured by the detector for a given position of the mirror (a given x value). That total intensity is the integral of the spectral intensity over all the range of spatial frequencies:

$$I_{meas}(x) = \int_{-\infty}^{\infty} I(x, \xi)d\xi = \int_{-\infty}^{\infty} I_0(\xi)[1 + \cos(2\pi\xi x)]d\xi$$

$$I_{meas}(x) = \int_{-\infty}^{\infty} I_0(\xi)d\xi + \int_{-\infty}^{\infty} I_0(\xi)\cos(2\pi\xi x)d\xi$$

$I_{meas}(x)$ is called the interferogram. The first integral on the right-hand side, $\int_{-\infty}^{\infty} I_0(\xi)d\xi$, is independent of the path difference x and represents the total intensity of the initial beam integrated over all frequencies (i.e., the area under the spectral intensity curve). This constant term is called the DC component, and we will define it as I_0. Thus:

$$I_{meas}(x) = I_0 + \int_{-\infty}^{\infty} I_0(\xi)\cos(2\pi\xi x)d\xi$$

By subtracting this DC term from the interferogram the oscillatory part is isolated:

$$I_{meas}(x) - I_0 = \int_{-\infty}^{\infty} I_0(\xi)\cos(2\pi\xi x)d\xi \qquad (8.3)$$

The difference $I_{meas}(x) - I_0$ is referred to as the normalized interferogram.

To understand the spectral reconstruction, recall the inverse Fourier transform definition:

$$f(x) = \int_{-\infty}^{\infty} F(\xi)\exp\left(i2\pi\xi x\right)\mathrm{d}\xi$$
$$= \int_{-\infty}^{\infty} F(\xi)\cos\left(2\pi\xi x\right)\mathrm{d}\xi + \int_{-\infty}^{\infty} F(\xi)i\,\sin(2\pi\xi x)\mathrm{d}\xi$$

(8.4)

When the function $F(\xi)$ is real and even, the product $F(\xi)\sin(2\pi\xi x)$ becomes odd, causing the second integral to vanish and the remaining expression reduce to a cosine transform. Since $I_0(\xi)$ satisfies these conditions (real and even), expression (8.3) demonstrate that the normalized interferogram represents the inverse Fourier transform of the spectrum $I_0(\xi)$. Therefore, the original spectrum can be recovered by applying the Fourier transform to the normalized interferogram.

So, the spectrum is obtained by calculating numerically the Fourier transform of the interferogram. This calculation is performed using the Fast Fourier Transform (FFT) algorithm. This calculation method will be outlined in a next paragraph.

FTIR has several advantages over traditional spectroscopy techniques. First, the entire spectrum is captured simultaneously, significantly speeding up data collection. Second, the measurement does not rely on narrow slits allowing the detector to receive much more light, which improves the signal-to-noise ratio. Finally, an interferometer is used instead of a monochromator, resulting in fewer movable parts and a less complex mechanical design.

8.2 Algorithms for the calculation of the Fourier transform

8.2.1 Sampling. Bandwidth. The Whittaker–Shannon theorem

Fourier transform numerical calculations are performed on sampled functions. In the context of optical images, the image is discretized and represented as an array of light intensity values. This is achieved by extracting intensity values from the function at specific spatial intervals (L) which correspond to a spatial frequency ξ_0. Intuitively, to accurately reproduce the image, a small spatial period (or equivalently a high spatial frequency) is desired. For example, abrupt changes in the intensity may be lost when the sampling frequency/period is too low/large. This is shown in figure 8.4 where the function $f(x)$ (plotted in red) is discretized using $L = 0.01$ (top) and $L = 0.1$ (bottom). As can be observed, several features of the function are lost in the last case due to the larger spatial period used for discretization, as evidenced by the curve plotted in black.

A more intriguing question is whether there is a fundamental limit to how small the sampling frequency can be and, if such a limit exists, what it is. This leads us to the Whittaker–Shannon theorem, a cornerstone in both optics and the broader field of converting analogue signals into digital form without losing information. The theorem has far-reaching applications, including signal processing, digital media transmission (such as music and audio), digital television, streaming, strioscopy, and image compression.

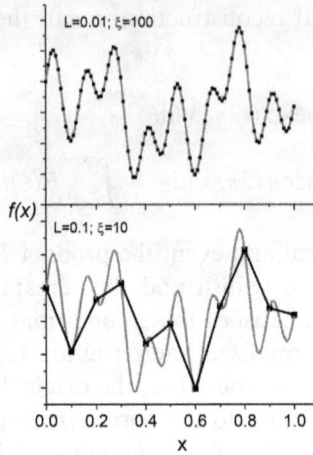

Figure 8.4. Discretization of a function $f(x)$ with a short (top) and large (bottom) spatial periods.

To illustrate the theorem and simplify its explanation, we will consider a 1D image whose intensity is described by a function $f(x)$. Let us multiply the function by a 1D periodic array of delta functions $\delta(x - nL)$, where $n = 0,\ 1,\ 2,\ \ldots$ and L is the spatial period. The result of this multiplication is an array of delta functions, where the area under each delta function is equal to the value of the function at $x = nL$. Mathematically, this can be expressed as:

$$f(x)\sum_{-\infty}^{\infty}\delta(x - nL) = \sum_{-\infty}^{\infty}f(nL)\delta(x - nL) \tag{8.5}$$

Taking the Fourier transform of this expression and using the corollary of the convolution theorem (equation (5.13)):

$$F\left\{f(x)\sum_{-\infty}^{\infty}\delta(x - nL)\right\} = F\{f(x)\} \otimes F\left\{\sum_{-\infty}^{\infty}\delta(x - nL)\right\} \tag{8.6}$$

Recall that the Fourier transform of a periodic array of delta functions is also a periodic array of delta functions in the frequency domain (equation (3.24)):

$$F\left\{\sum_{-\infty}^{\infty}\delta(x - nL)\right\} = \frac{1}{L}\sum_{-\infty}^{\infty}\delta(\xi - n\xi_0) \tag{8.7}$$

where $\xi_0 = \frac{1}{L}$ is the sampling frequency. Substituting equation (8.7) in (8.6), we get:

$$F\left\{f(x)\sum_{-\infty}^{\infty}\delta(x - nL)\right\} = F\{f(x)\} \otimes \frac{1}{L}\sum_{-\infty}^{\infty}\delta(\xi - n\xi_0) \tag{8.8}$$

From equation (8.8), it can be observed, based on the properties of the convolution of a function with a delta function (equations (5.9) and (5.10)), that the Fourier transform of equation (8.5) corresponds to the Fourier transform of $f(x)$ replicated at the positions of the delta functions in the summation. Consequently, the Fourier transform of a sampled function becomes a periodic function in the frequency domain.

Now, let us consider that the Fourier transform of $f(x)$ is zero outside a specific finite range of frequencies $[\xi_{min}, \xi_{max}]$. In order to simplify the analysis, let us consider the case where Fourier transform is even, so that $\xi_{min} = -\xi_{max}$. In other words, the function $f(x)$ contains no frequency components higher than a specific cutoff frequency modulus ξ_{max}. Such functions are referred to as 'band-limited functions'. Then, for band-limited functions:

$$|\xi| \leqslant \xi_{max}$$

In figures 8.5, panels (a) and (b) depict a hypothetical band-limited function and its Fourier transform, respectively. Panels (c) and (d) show the periodic array of delta functions and its Fourier transform, respectively. Finally, panels (e) and (f), present the result of the product in equation (8.5) and its Fourier transform (equation (8.8)), respectively. As observed, the Fourier transform of equation (8.5) results in a periodic version of the Fourier transform of the function $f(x)$, with periodicity determined by $\xi_0 = \frac{1}{L}$.

To exactly reconstruct the original function $f(x)$, it is necessary to eliminate frequencies higher than $|\xi_{max}|$, retain only the term with $n = 0$, and perform the

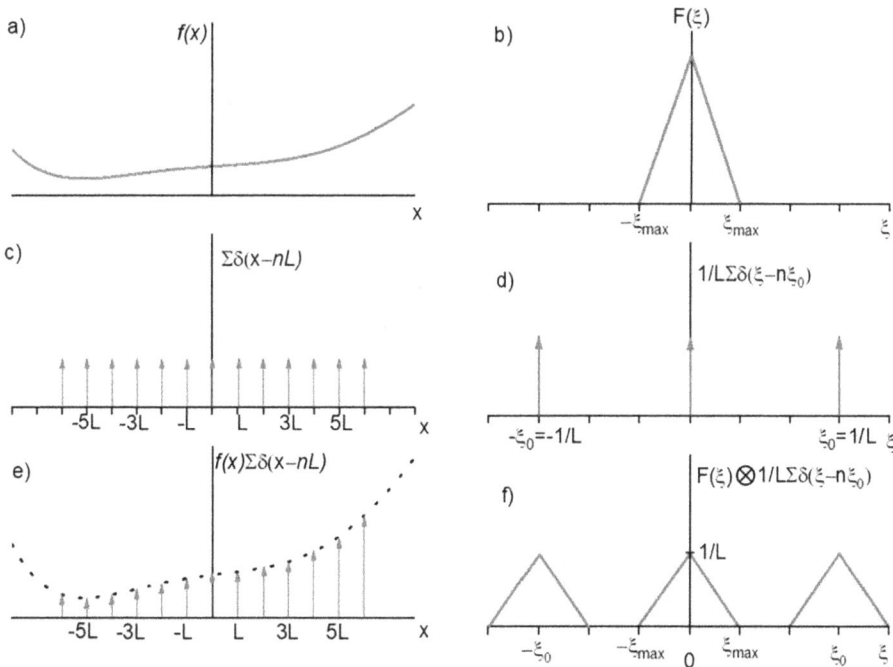

Figure 8.5. (a, b) The function $f(x)$ and its Fourier transform $F(\xi)$; (c, d) the periodic array of delta functions and its Fourier transform; and (e, f) $f(x)\sum_{-\infty}^{\infty}\delta(x - nL)$ and its Fourier transform.

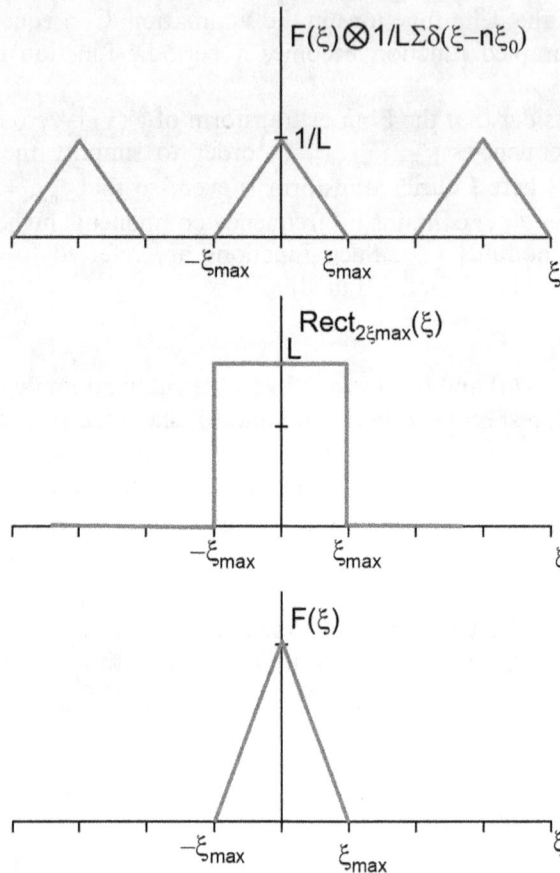

Figure 8.6. The periodic Fourier transform of figure 8.5(f) (top) is multiplied by a rectangular function $L \, \text{Rect}_{2\xi_{max}}(\xi)$ (middle). The result (bottom) is the Fourier transform (figure 8.5(b)) of the function $f(x)$. Applying the inverse Fourier transform, the function $f(x)$ can be retrieved.

inverse Fourier transform. To remove these higher frequencies, equation (8.8) must be multiplied by a rectangular function (equation (3.11)) in the frequency domain, $L \, \text{Rect}_{2\xi_{max}}(\xi)$ as illustrated in figure 8.6.

For this procedure to work correctly, the sampling frequency ξ_0 must be greater than or equal to $2\xi_{max}$. As illustrated in figure 8.7, if the sampling frequency is lower than $2\xi_{max}$, the multiplication by $\text{Rect}_{2\xi_{max}}(\xi)$ will not retrieve the Fourier transform of $f(x)$, but a distorted function whose inverse Fourier transform will not reproduce adequately $f(x)$.

This phenomenon is called aliasing. The critical frequency is called Nyquist frequency (in honour to the Swedish American physicist and electronic engineer Harry Nyquist, 1889–1976) and is defined as:

$$\xi_{\text{Nyquist}} = 2\xi_{max}.$$

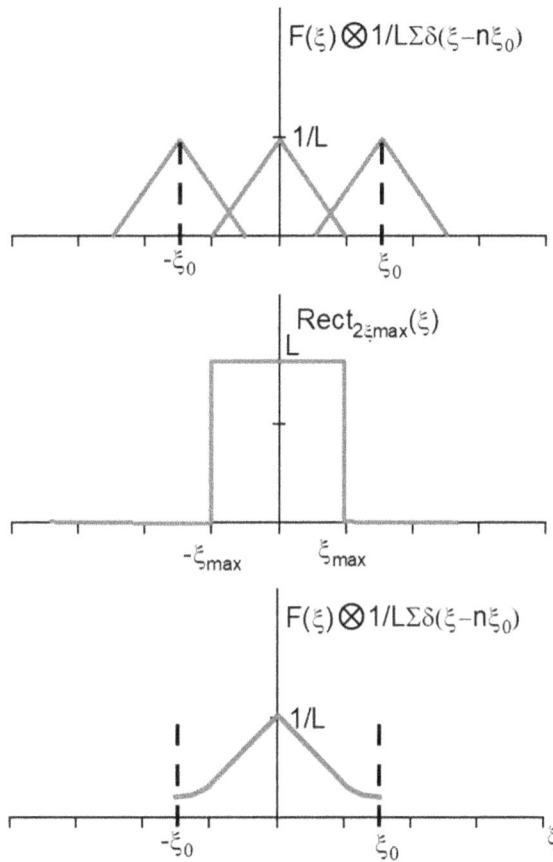

Figure 8.7. The case in which the sampling frequency ξ_0 is smaller than $2\xi_{max}$. Multiplying this function for the $L \, \mathrm{Rect}_{2\xi_{max}}(\xi)$ as in figure 8.6 will produce a distorted version of the function $f(x)$.

'For a function to be properly recovered from its sampled version, the sampling rate must be at least twice the frequency of the highest frequency component of the function'. This assertion is known as the Whittaker–Shannon or Nyquist. Shannon theorem.

8.2.2 The discrete Fourier transform

In practice, the calculation of Fourier transforms is carried out using numerical methods. The general algorithm is the so-called Discrete Fourier Transform (DFT) which will be outlined in what follows. The function $f(x)$ is discretized and converted in a list of values which can be retrieved from the integration of equation (8.5) and using the sifting property of the delta functions (see equation (3.19)). So:

$$\int_{-\infty}^{\infty} f(x) \sum_{-\infty}^{\infty} \delta(x - nL) \mathrm{d}x = \sum_{-\infty}^{\infty} f(nL) \tag{8.9}$$

Then, the Fourier transform of the sampled function can be written as:

$$F(\xi) = \sum_{-\infty}^{\infty} f_S(nL)\exp(-i2\pi\xi nL) \tag{8.10}$$

Since it is impossible to store an infinite amount of data on a computer, the series must be truncated to have a finite number of N terms. Then, equation (8.10) becomes:

$$F(\xi) = \sum_{n=0}^{N-1} f_S(n)\exp(-i2\pi\xi nL) \tag{8.11}$$

in which the argument of the function was written as n instead of nL considering that L is a constant. In the scope of building a discrete equation for the Fourier transform, the frequency must be discretized as well. As was demonstrated in the precedent section, the Fourier transform of the sampled function is a periodic function in the frequency domain with period equal to $\xi_0 = \frac{1}{L}$. To discretize the frequency, we will divide ξ_0 in $M\,\Delta\xi_0$ finite elements in such a way that $\Delta\xi_0 = \frac{\xi_0}{M}$ and for simplicity we will make $M = N$. The variable ξ will become $k\Delta\xi_0 = k\frac{\xi_0}{N}$. Substituting this into equation (8.11), we obtain:

$$F(k) = \sum_{n=0}^{N-1} f_S(n)\exp\left(-i2\pi k\frac{L\xi_0}{N}n\right)$$

As $\xi_0 = \frac{1}{L}$, this simplifies to:

$$F(k) = \sum_{0}^{N-1} f_S(n)\exp\left(-i\frac{2\pi}{N}kn\right) \tag{8.12}$$

This is the DFT equation.

As an example, let's consider a function with $N = 4$, and $f_S(n) = \{0,\ 3,\ -4,\ 2\}$. Then, according to equation (8.12):

$$F(0) = f_S(0)\exp\left\{-i\frac{2\pi}{4}(0)(0)\right\} + f_S(1)\exp\left\{-i\frac{2\pi}{4}(0)(1)\right\}$$
$$+ f_S(2)\exp\left\{-i\frac{2\pi}{4}(0)(2)\right\} + f_S(3)\exp\left\{-i\frac{2\pi}{4}(0)(3)\right\}$$
$$= 0 + 3 - 4 + 2 = 1$$

$$F(1) = f_S(0)\exp\left\{-i\frac{2\pi}{4}(1)(0)\right\} + f_S(1)\exp\left\{-i\frac{2\pi}{4}(1)(1)\right\}$$
$$+ f_S(2)\exp\left\{-i\frac{2\pi}{4}(1)(2)\right\} + f_S(3)\exp\left\{-i\frac{2\pi}{4}(1)(3)\right\}$$
$$= 0 + 3\exp\left(-i\frac{\pi}{2}\right) - 4\exp(-i\pi) + 2\exp\left(-i\frac{3\pi}{2}\right) = 0 - 3i + 4 + 2i = 4 - i$$

$$F(2) = f_S(0)\exp\left\{-i\frac{2\pi}{4}(2)(0)\right\} + f_S(1)\exp\left\{-i\frac{2\pi}{4}(2)(1)\right\}$$
$$+ f_S(2)\exp\left\{-i\frac{2\pi}{4}(2)(2)\right\} + f_S(3)\exp\left\{-i\frac{2\pi}{4}(2)(3)\right\}$$
$$= 0 + 3\exp(-i\pi) - 4\exp(-i2\pi) + 2\exp(-i3\pi) = 0 - 3 - 4 - 2 = -9$$

$$F(3) = f_S(0) \exp\left\{-i\frac{2\pi}{4}(3)(0)\right\} + f_S(1) \exp\left\{-i\frac{2\pi}{4}(3)(1)\right\}$$

$$+ f_S(2) \exp\left\{-i\frac{2\pi}{4}(3)(2)\right\} + f_S(3) \exp\left\{-i\frac{2\pi}{4}(3)(3)\right\}$$

$$= 0 + 3\exp\left(-i\frac{3\pi}{2}\right) - 4\exp(-i3\pi) + 2\exp\left(-i\frac{9\pi}{2}\right) = 0 + 3i + 4 - 2i = 4 + i$$

$$F(4) = f_S(0) \exp\left\{-i\frac{2\pi}{4}(4)(0)\right\} + f_S(1) \exp\left\{-i\frac{2\pi}{4}(4)(1)\right\}$$

$$+ f_S(2) \exp\left\{-i\frac{2\pi}{4}(4)(2)\right\} + f_S(3) \exp\left\{-i\frac{2\pi}{4}(4)(3)\right\}$$

$$= 0 + 3\exp(-i2\pi) - 4\exp(-i4\pi) + 2\exp(-i6\pi) = 0 + 3 - 4 + 2 = 1$$

Thus, the Fourier transform of the sampled function $f_S(n) = \{0, 3, -4, 2\}$ in the space domain, is the list $F(k) = \{1, 4 - i, -9, 4 + i\}$ in the spatial frequency domain. Note that $F(4) = F(0)$ as expected since the Fourier transform is periodic in the frequency domain with period N and $N = 4$.

Equation (8.12) can be modified making $\exp\left(-i\frac{2\pi}{N}\right) = W_N$ which is called the twiddle factor. Then, equation (8.12) becomes:

$$F(k) = \sum_{0}^{N-1} f_S(n) W_N^{kn} \tag{8.13}$$

This equation defines $F(k)$ as a vector which is the product of the matrix W_N^{kn} multiplied by a vector $f_S(n)$:

$$
\begin{vmatrix} F(0) \\ F(1) \\ F(2) \\ F(3) \end{vmatrix}
=
\begin{vmatrix}
W_N^{00} & W_N^{01} & W_N^{02} & W_N^{03} \\
W_N^{10} & W_N^{11} & W_N^{12} & W_N^{13} \\
W_N^{20} & W_N^{21} & W_N^{22} & W_N^{23} \\
W_N^{30} & W_N^{31} & W_N^{32} & W_N^{33}
\end{vmatrix}
\begin{vmatrix} f_S(0) \\ f_S(1) \\ f_S(2) \\ f_S(3) \end{vmatrix}
$$

Since $W_N^{oo} = W_N^{01} = W_N^{02} = W_N^{03} = W_N^{10} = W_N^{20} = W_N^{30} = 1$, the matrix simplifies to:

$$
\begin{vmatrix} F(0) \\ F(1) \\ F(2) \\ F(3) \end{vmatrix}
=
\begin{vmatrix}
1 & 1 & 1 & 1 \\
1 & W_N^{11} & W_N^{12} & W_N^{13} \\
1 & W_N^{21} & W_N^{22} & W_N^{23} \\
1 & W_N^{31} & W_N^{32} & W_N^{33}
\end{vmatrix}
\begin{vmatrix} f_S(0) \\ f_S(1) \\ f_S(2) \\ f_S(3) \end{vmatrix}
$$

It can be noted that the exponent kn represent a multiplication of the numbers k and n, not an index. In our example, we had to perform four multiplications for every one of the four components of the Fourier transform, in total $4 \times 4 = 16$ multiplications, or N^2. Considering the multiplication as the key operation, the complexity (the amount of time the algorithm takes to complete) increases with the number of samples as N^2. In practice, real sampled functions contain much more than four elements, frequently powers of 2 (256, 1024, etc) and in those cases the number of multiplications is quite large, and the computation very time-consuming. For N samples, the matrix equation becomes:

$$\begin{vmatrix} F(0) \\ F(1) \\ F(2) \\ F(3) \\ \vdots \\ F(N-1) \end{vmatrix} = \begin{vmatrix} 1 & 1 & 1 & 1 & \cdot \cdot & 1 \\ 1 & W_N^{11} & W_N^{12} & W_N^{13} & \cdot \cdot & W_N^{1(N-1)} \\ 1 & W_N^{21} & W_N^{22} & W_N^{23} & \cdot \cdot & W_N^{2(N-1)} \\ 1 & W_N^{31} & W_N^{32} & W_N^{k33} & \cdot \cdot & W_N^{3(N-1)} \\ \vdots & \vdots & \vdots & \vdots & \cdot \cdot & \vdots \\ 1 & W_N^{(N-1)1} & W_N^{(N-1)2} & W_N^{(N-1)3} & \cdot \cdot & W_N^{(N-1)(N-1)} \end{vmatrix} \begin{vmatrix} f_S(0) \\ f_S(1) \\ f_S(2) \\ f_S(3) \\ \vdots \\ f(N-1) \end{vmatrix}$$

In the next section, an algorithm to make these calculations much more efficient will be explained: the Fast Fourier Transform (FFT).

8.2.3 The Fast Fourier transform

To reduce the complexity of the DFT algorithm, which results in very large execution times and memory requirements, an important algorithm was developed by James W Cooley and John W Tukey in 1965: the Fast Fourier Transform (FFT). This algorithm takes advantage of the 'divide and conquer' approach. Specifically, the N-sample summation in equation (8.12) or (8.13) is successively divided into smaller summations until only two-sample summations remain. To achieve this, the even and odd terms of $f_S(n)$ are separated into two distinct summations. Thus:

$$F(k) = \sum_{n=0}^{N-1} f_S(n)\exp(-i\tfrac{2\pi}{N}kn)$$

$$= \sum_{m=0}^{\frac{N}{2}-1} f_S(2m)\exp(-i\tfrac{2\pi}{N}k2m) + \sum_{m=0}^{\frac{N}{2}-1} f_S(2m+1)\exp\left\{-i\tfrac{2\pi}{N}k(2m+1)\right\}$$

Here, the variable n was replaced by m for separating the summations.

The equation can be rewritten as:

$$F(k) = \sum_{m=0}^{\frac{N}{2}-1} f_S(2m)\exp(-i\tfrac{2\pi}{N}k2m)$$

$$+ \exp(-i\tfrac{2\pi}{N}k) \sum_{m=0}^{\frac{N}{2}-1} f_S(2m+1)\exp\left\{-i\tfrac{2\pi}{N}k(2m)\right\}$$

Note that $\exp(-i\tfrac{2\pi}{N}k2m) = \exp(-i\tfrac{2\pi}{N/2}km)$, we obtain:

$$F(k) = \sum_{m=0}^{\frac{N}{2}-1} f_S(2m)\exp(-i\tfrac{2\pi}{N/2}km)$$

$$+ \exp(-i\tfrac{2\pi}{N}k) \sum_{m=0}^{\frac{N}{2}-1} f_S(2m+1)\exp(-i\tfrac{2\pi}{N/2}km)$$

Or using the twiddle factor $W_N = \exp(-i\tfrac{2\pi}{N})$ defined earlier, this becomes:

$$F(k) = \sum_{m=0}^{\frac{N}{2}-1} f_S(2m) W_{N/2}^{km} + W_N^k \sum_{m=0}^{\frac{N}{2}-1} f_S(2m+1) W_{N/2}^{km} \tag{8.14}$$

It is important to note that k runs from $k = 0$ to $k = N - 1$ ($F(k)$ is a list of N samples) while the summations on the right-hand side from $m = 0$ to $m = \frac{N}{2} - 1$ (they are two lists of $\frac{N}{2}$ samples). From equation (8.12) we recognize these summations as the DFTs of the even and odd components of f_S, respectively.

Let us denote the summation corresponding to the even components of f_S as $G(k)$ and the summation corresponding with the odd components of f_S as $H(k)$. Equation (8.14) can be written as:

$$F(k) = G(k) + W_N^k H(k)$$

where $G(k)$ and $H(k)$ are both Fourier transforms of lists of $\frac{N}{2}$ samples. As demonstrated earlier (see section 8.2.1), the Fourier transform is periodic in the frequency domain with periodicity equal to the sampling frequency, (that is, with the number of samples), then, $G(k)$ and $H(k)$ are periodic with period $N/2$. Therefore, when evaluating them in the range:

$$0 \leqslant k \leqslant \left(\frac{N}{2} - 1\right)$$

the same values are obtained as when k is evaluated in the range:

$$\frac{N}{2} \leqslant k \leqslant (N - 1)$$

Equivalently:

$$G(k - N/2) = G(k)$$

And

$$H(k - N/2) = H(k)$$

These quantities need to be evaluated only once instead of twice.

However, the term W_N^k is not the same for both ranges. In the second range:

$$W_N^{k+N/2} = \exp\left\{-i\frac{2\pi}{N}(k + N/2)\right\} = \exp\left(-i\frac{2\pi}{N}k\right)\exp\left\{-i\frac{2\pi}{N}(N/2)\right\}$$
$$= \exp\left(-i\frac{2\pi}{N}k\right)\exp(-i\pi) = -\exp\left(-i\frac{2\pi}{N}k\right) = -W_N^k \tag{8.15}$$

Then,

$$F(k) = G(k) + W_N^k H(k)$$

In the range $0 \leqslant k \leqslant (\frac{N}{2} - 1)$, and:

And,

$$F(k) = G(k - N/2) - W_N^{k-N/2}H(k - N/2)$$

In the range $\frac{N}{2} \leqslant k \leqslant (N-1)$

Then, we can express $F(k)$ as shown in table 8.1.

Notably, with the values of $G(k)$ and $H(k)$ for $k = 0, 1, 2, \ldots, \frac{N}{2} - 1$, all N values of $F(k)$ can be determined. For example, using $G(0)$ and $H(0)$, we can compute the term $k = 0$:

$$F(0) = G(0) + W_N^0 H(0)$$

and the term $k = 0 + \frac{N}{2}$:

$$F\left(0 + \frac{N}{2}\right) \equiv F\left(\frac{N}{2}\right) = G(0) - W_N^0 H(0)$$

This process is illustrated using a graphical representation, called 'a butterfly diagram', as shown in figure 8.8. The meaning of the figure is as follows: the value of $F(0)$ is formed by the sum of $G(0)$ (red line) and $H(0)$ multiplied by W_N^0 (blue line with the indication W_N^0). Similarly, the value of $F(\frac{N}{2})$ is formed by the sum of $G(0)$ (green line) and $H(0)$ multiplied by $-W_N^0$ (purple line with the indication $-W_N^0$).

Table 8.1. Components of $F(k)$ as calculated from $G(k)$, $H(k)$ and the twiddle factor.

k	$G(k)$	$H(k)$	$F(k)$
0	$G(0)$	$H(0)$	$F(0) = G(0) + W_N^0 H(0)$
1	$G(1)$	$H(1)$	$F(1) = G(1) + W_N^1 H(1)$
2	$G(2)$	$H(2)$	$F(2) = G(2) + W_N^2 H(2)$
.	.	.	.
$\frac{N}{2} - 1$	$G(\frac{N}{2} - 1)$	$H(\frac{N}{2} - 1)$	$F(\frac{N}{2} - 1) = G(\frac{N}{2} - 1) + W_N^{\frac{N}{2}-1} H(\frac{N}{2} - 1)$
$\frac{N}{2}$	$G(0)$	$H(0)$	$F(\frac{N}{2}) = G(0) - W_N^0 H(0)$
$\frac{N}{2} + 1$	$G(1)$	$H(1)$	$F(\frac{N}{2} + 1) = G(1) - W_N^1 H(1)$
$\frac{N}{2} + 2$	$G(2)$	$H(2)$	$F(\frac{N}{2} + 2) = G(2) - W_N^2 H(2)$
.	.	.	.
$N - 1$	$G(\frac{N}{2} - 1)$	$H(\frac{N}{2} - 1)$	$F(N - 1) = G(\frac{N}{2} - 1) - W_N^{\frac{N}{2}-1} H(\frac{N}{2} - 1)$

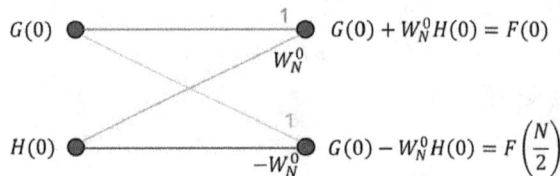

Figure 8.8. 'Butterfly diagram' representation of the formation of the elements $F(0)$ and $F(\frac{N}{2})$ from the elements $G(0)$ and $H(0)$ taken from the DFTs of the even and odd lists, respectively.

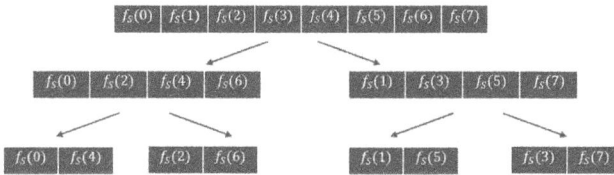

Figure 8.9. Decimation process for a sampled function with $N = 8$.

In one of the most common ways of implementing the FFT algorithm, called decimation, the sample lists are successively divided in two, always separating the even and odd samples. Starting from the original list of N samples, two lists of $\frac{N}{2}$ samples each are formed. These are then further divided into four lists of $\frac{N}{4}$ samples, which in turn are divided into eight lists of $\frac{N}{8}$ samples, and so on, until only lists of two samples remain.

It can be observed that the number of lists in each step is 2^γ where γ is the step number (e.g. two lists in the first step, four in the second, eight in the third and so on). Consequently, the number of samples in each list is equal to the total number of samples N divided by the number of lists, that is, $\frac{N}{2^\gamma}$. To reach lists of two-samples, we set $\frac{N}{2^\gamma} = 2$ which implies that the number of steps is $\gamma = \log_2 N - 1$. Since $N \gg 1$ in most practical cases, the number of steps is approximately $\log_2 N$.

In figure 8.9, the process of decimation is illustrated for a hypothetical $f_S(n)$ with $N = 8$. The eight-samples list is divided in two: one containing the even-indexed samples and the other containing the odd-indexed samples. Each of these two lists is then further divided into two using the same procedure. In this case, after only three steps, lists of two samples are obtained. In fact, since $N = 8$, the total number of steps γ is $\log_2 N = 3$. The FFT calculation process begins with the two-samples lists (the bottom row in figure 8.9) and progresses step by step to reach $F(k)$.

To determine the distribution of samples in the bottom line (which, for N large, is not as trivial as in this simple case) the 'bit reversal' procedure can be applied. This involves expressing the index of each sample in binary representation and then mapping each index to its bit-reversed counterpart. For $N = 8$, the process is as follows:

n	In binary	Reversed	Mapped
0	000	000	0
1	001	100	4
2	010	010	2
3	011	110	6
4	100	001	1
5	101	101	5
6	110	011	3
7	111	111	7

Note that the mapped indices in the last column correspond to the order of samples in the bottom row of figure 8.9. The FFT calculation begins from this bottom row, by computing the Fourier transform of these two-sample lists. It is important to note that the Fourier transform of the first two-sample list $\{f_s(0), f_s(4)\}$ represents the Fourier transform of the even-indexed elements of the list to the left in the next step. This Fourier transform corresponds to the $G(0)$, which is necessary for that calculation step. Similarly, the Fourier transform of the second two-sample list $\{f_s(2), f_s(6)\}$ is $H(0)$. The third and fourth two-sample lists correspond to the $G(0)$ and $H(0)$, respectively, and serve the same purpose for the list on the right in the next step.

8.2.3.1 Example of the FFT algorithm with a $N = 8$ list

Next, we will illustrate how the FFT operates in a $N = 8$ list. However, we will first consider the case of a two-sample list, as this is the foundational step in the FFT procedure. For a list of $N = 2$ composed of $f_S(0)$ and $f_S(1)$, the Fourier transform, according to equation (8.13), will be:

$$F(k) = \sum_0^1 f_S(n) W_2^{kn}$$

Then,

$$F(0) = f_S(0) W_2^{00} + f_S(1) W_2^{01}$$

And

$$F(1) = f_S(0) W_2^{10} + f_S(1) W_2^{11}$$

As $W_2^{00} = W_2^{01} = W_2^{10} = 1$ and $W_2^{11} = -1$

$$F(0) = f_S(0) + f_S(1)$$

$$F(1) = f_S(0) - f_S(1)$$

The corresponding reasoning using the butterfly representation is shown in figure 8.10.

Let us consider now the eight-samples list which represents the function $f_S(n)$ and $f_S(n) = \{1, \ 3, \ 0, \ 2, \ 5, \ 1, \ 3, \ 1\}$.

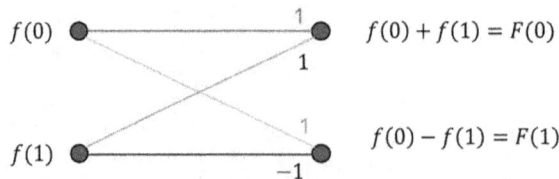

Figure 8.10. The butterfly representation for a two-samples list.

Doing the bit reversal procedure, our starting list for the FFT calculation will be:

$$\{1, \ 5, \ 0, \ 3, \ 3, \ 1, \ 2, \ 1\}$$

The starting line is shown in figure 8.11:

Now we have to resolve the Fourier transform of each couple using the butterfly procedure as shown in figure 8.12:

Note that the list $\{6, -4\}$ is the Fourier transform of $\{f_s(0), f_s(4)\}$, the even-indexed elements of the next list according to Figure 8.9, and the list $\{3, -3\}$ is the Fourier transform of $\{f_s(2), f_s(6)\}$, the odds-indexed elements of the next list in figure 8.9. Similarly, $\{4, 2\}$ and $\{3, 1\}$ are the Fourier transform of the even- and odds-indexed elements, respectively, of the other list in the next step.

The next step is to calculate the Fourier Transform of the four samples summations by using the previous result and the twiddle factor that for this case is $W_4^k = \exp\left(-i\frac{2\pi}{4}k\right)$. Now $W_4^0 = 1$ and $W_4^1 = -i$. The result of the Fourier transform up to this step is shown at the end of the butterflies (figure 8.13).

The final step is to obtain the definitive $F(k)$ from the two four-sample lists from the previous result and using the twiddle factor $W_8^k = \exp\left(-i\frac{2\pi}{8}k\right)$ which is 1, $\frac{\sqrt{2}}{2} - \frac{\sqrt{2}}{2}i, -i, -\frac{\sqrt{2}}{2} - \frac{\sqrt{2}}{2}i$, for $k = 0, 1, 2, 3$, respectively. It can be noted that the first four-sample summation corresponds to the Fourier transform of the even elements of the final list, and the second four-sample summation to the odd elements. The calculation of this step is presented in figure 8.14.

Thus:

$$F(k) = 16, \ -4 + 3i + \frac{1 - 3i}{\sqrt{2}}, \ 3 - i, \ -4 - 3i - \frac{1 + 3i}{\sqrt{2}}, \ 2,$$
$$-4 + 3i - \frac{1 - 3i}{\sqrt{2}}, \ 3 + i, \ -4 - 3i + \frac{1 + 3i}{\sqrt{2}}.$$

$$f_S(0) = 1 \quad \bullet$$

$$f_S(4) = 5 \quad \bullet$$

$$f_S(2) = 0 \quad \bullet$$

$$f_S(6) = 3 \quad \bullet$$

$$f_S(1) = 3 \quad \bullet$$

$$f_S(5) = 1 \quad \bullet$$

$$f_S(3) = 2 \quad \bullet$$

$$f_S(7) = 1 \quad \bullet$$

Figure 8.11. The starting line for the calculation of the FFT of the eight-samples list.

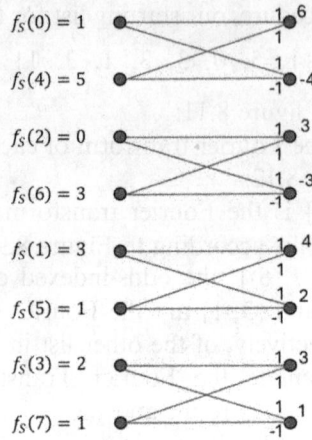

Figure 8.12. The first step of the FFT calculation.

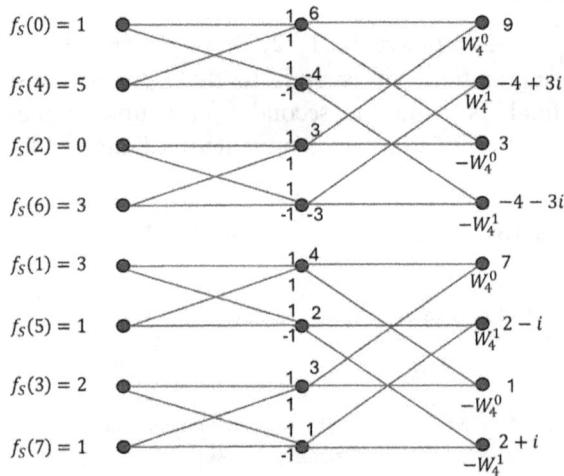

Figure 8.13. The second step of the FFT calculation of the eight-samples list.

8.2.3.2 The complexity of the FFT

It can be noted that in each step there are $\frac{N}{2}$ butterflies, and in every butterfly only one of the couples is multiplied by the twiddle factor, resulting in one complex multiplication for every butterfly. Since the number of steps in the FFT algorithm is $\log_2 N$, the total number of multiplications is $\frac{N}{2}\log_2 N$. Additionally, there are complex additions in the FFT algorithm. Each butterfly involves two complex additions, so the total number of complex additions is $N\log_2 N$. This can be compared to the standard DFT algorithm, which requires N^2 complex multiplications and $N(N-1)$ complex additions (since there are $N-1$ additions for each

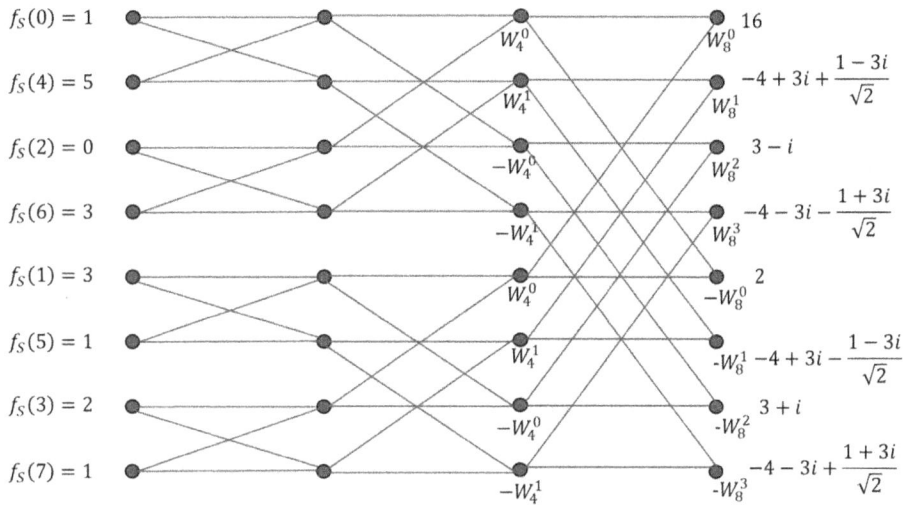

Figure 8.14. The third step of the FFT calculation of the eight-samples list.

of the N bins k). Since multiplications are generally more computationally expensive than additions, the computational complexity of the FFT is often considered to be dominated by the number of multiplications, which is $\frac{N}{2}\log_2 N$. For example, for a list of 1024 samples, the number of complex multiplications in DFT is 1 048 576 while the FFT algorithm requires only 5120.

8.3 Holography

Holography is an imaging process that allows one to record and visualize scenes or objects in three dimensions [1, 2] instead of two as is the case in photography. On a photograph, each individual point from the scene governs the intensity at a single position in the image plane, as we can see in figure 2.7, which thus is saved onto one individual pixel of the camera. Every pixel detects only the light field intensity, as we saw in section 2.4.1, and not the phase. Then when one looks at the photograph, the phase information cannot be retrieved, which means we cannot know how far each object point was from the camera. In 1948, Dennis Gabor figured that a way to record both the intensity and the phase at once is to interfere the light reflected off every point of the scene we want to record with a reference beam. The idea of recording it all (intensity and phase) is encrypted into the word 'holography' as 'holos' means 'whole' and 'graphein' means 'to write'. We have seen in section 2.4.3 that in order to observe interferences, it is necessary that the light beams (the reference one and the one illuminating the recorded scene) are coherent, which was enabled only in 1960 by the invention of the laser.

8.3.1 General principle

The process is constituted of two steps, the recording and the reading (or reconstruction).

8.3.1.1 Recording

Let us write the wavefront spatial complex amplitude that is reflected off the coherently illuminated scene as:

$$\underline{\epsilon}(\vec{r}) = \epsilon_0(\vec{r}) \exp(-i\Phi(\vec{r})) \text{ where } \Phi(\vec{r}) = \vec{k}_s \cdot \vec{r} + \phi_0$$

and the reference beam amplitude as:

$$\underline{\mathcal{E}}(\vec{r}) = \mathcal{E}_0(\vec{r}) \exp(-i\Psi(\vec{r})) \text{ where } \Psi(\vec{r}) = \vec{k}_{\text{ref}} \cdot \vec{r} + \psi_0$$

Note that $\vec{k}_s$ depends on $\vec{r}$ and $\vec{k}_{\text{ref}}$ can depend on $\vec{r}$ as well, unless it is a plane wave.

The intensity distribution obtained by interfering these two fields is given by equation (2.8), which here gives:

$$I(\vec{r}) = \frac{K}{2}(\epsilon_0(\vec{r})^2 + \mathcal{E}_0(\vec{r})^2 + 2\epsilon_0(\vec{r})\mathcal{E}_0(\vec{r})\cos(\Phi(\vec{r}) - \Psi(\vec{r})))$$

This expression shows that the intensity distribution contains information about the intensities of the reference and reflected wavefronts, as well as their phases difference. That interference pattern has to be recorded on a material that can detect intensity variations with a high resolution to keep all the necessary information from the scene to be recorded. Also, the dependency of the interferogram (here called a hologram) transmission must be linear with the intensity it is exposed to. This is the case for some range of intensity values for the usual materials, where one can write $T(\vec{r}) = \alpha I(\vec{r})$ where α depends on the exposure time of the hologram. A very commonly used material is the photographic plate, that is constituted of a glass plate coated with silver salts, in which the silver ions Ag^{2+} are turned into metallic silver (Ag) under the action of light. It is also possible to electronically detect holograms, or even generate them using computers.

The hologram transmission function (see section 6.1) is thus given by:

$$T(\vec{r}) = \frac{\alpha K}{2}(|\underline{\mathcal{E}}(\vec{r})|^2 + |\underline{\epsilon}(\vec{r})|^2 + \underline{\mathcal{E}}(\vec{r})\underline{\epsilon}^*(\vec{r}) + \underline{\mathcal{E}}^*(\vec{r})\underline{\epsilon}(\vec{r})). \tag{8.16}$$

8.3.1.2 Reconstruction

The reconstruction process consists in illuminating the hologram with a wavefront that is identical to the reference beam: $\underline{\mathcal{E}}(\vec{r})$. The transmitted wave $\underline{\mathcal{E}}_t(\vec{r})$ after the mask has the following expression:

$$\underline{\mathcal{E}}_t(\vec{r}) = T(\vec{r})\underline{\mathcal{E}}(\vec{r}) \propto (|\underline{\mathcal{E}}|^2 + |\underline{\epsilon}|^2)\underline{\mathcal{E}} + \underline{\mathcal{E}}\,\underline{\epsilon}^*\,\underline{\mathcal{E}} + \underline{\mathcal{E}}^*\,\underline{\epsilon}\,\underline{\mathcal{E}} \tag{8.17}$$

where we dropped the $(\vec{r})$ dependencies for the sake of readability. The first term is simply proportional to the reference beam field. The second term is equal to:

$$\underline{\mathcal{E}}^2\underline{\epsilon}^* = \mathcal{E}_0^2\epsilon_0 \exp(-i(2\Psi - \Phi))$$

whose propagation direction is ruled by the vector $2\vec{k}_{\text{ref}} - \vec{k}_s$. Finally, the most interesting term is the third one, as it is equal to:

$$\underline{\mathcal{E}}^*\,\underline{\epsilon}\,\underline{\mathcal{E}} = \mathcal{E}_0^2\epsilon_0 \exp(-i\Phi) = \mathcal{E}_0^2\underline{\epsilon}$$

and thus accurately represents the field reflected off the scene during the recording step of the process.

8.3.1.3 Tilting of the reference beam

In the case where the reference beam is chosen to be normal to the holographic plate, the field after the holographic plate during the reconstruction step is given by the Fourier transform of the transmission function (8.16). This Fourier transform includes three components: a Dirac function in the optical axis direction, and two more Dirac functions corresponding to the Fourier transform of the cosine contribution. One arises in the $\vec{k}_s$ direction and the other in the $2\vec{k}_{\text{ref}} - \vec{k}_s$ direction. But then the propagation direction of the reconstructed scene beam (third term in equation (8.17)) and that of the reference beam (first term in equation (8.17)) overlap, which prevents the viewer from seeing only the reconstructed scene. This is why we usually tilt the reference beam with respect to the optical axis (figure 8.15).

8.3.2 Example of holography for a single point in space

Let us consider a simple case that offers an intuitive understanding of how holography works. We examine a single point-like object in space, illuminated by a reference beam. This simplified scenario is pedagogically valuable, because it makes it possible to develop the full calculations involved in the recording and reading process of holography. This approach was inspired by [3].

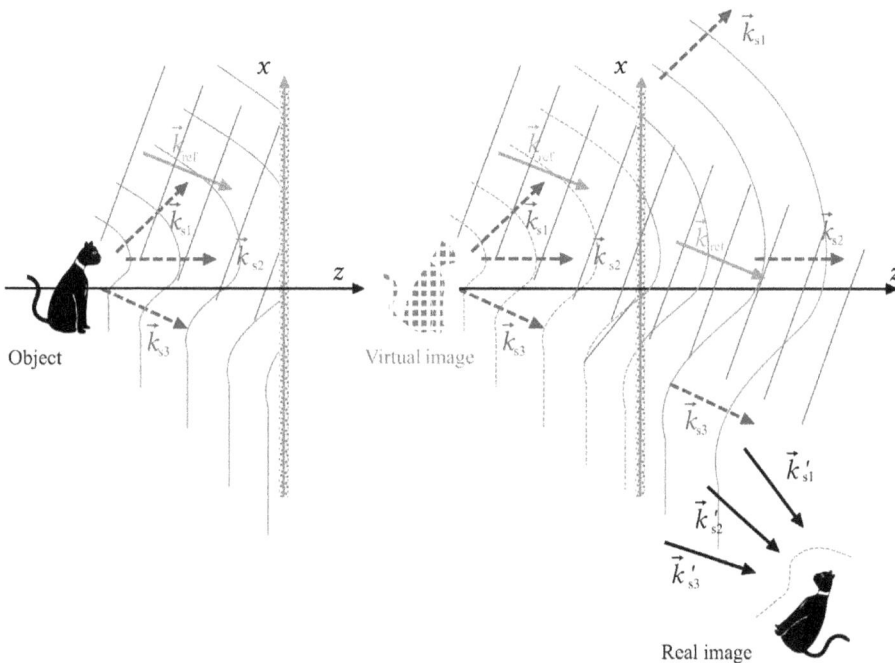

Figure 8.15. Principle of holography recording (left panel) and holography reconstruction (right panel). The vectors $\vec{k}'_{s1}$, $\vec{k}'_{s2}$ and $\vec{k}'_{s3}$ correspond to the reconstructed scene beams. If the reference beam were not tilted, it would overlap with $\vec{k}'_{s2}$, and the viewer would not be able to see the image of the cat.

8.3.2.1 Recording of the hologram

The point re-emits the light as a spherical wave, as represented in figure 2.10. The spherical light field interferes with the reference beam on a film. The intensity on the film is given by the formula of interferences (equation (2.7)):

$$I_{\text{point}+\text{ref}} = \frac{K}{2}\left(E_{\text{point}}^2 + E_{\text{ref}}^2 + 2E_{\text{point}}E_{\text{ref}}\cos\left(\frac{2\pi}{\lambda_0}\Delta\delta\right)\right) \qquad (8.18)$$

where $\Delta\delta$ can be calculated from the 2D drawing in figure 8.16. We have:

$$\Delta\delta = \sqrt{D^2 + x^2 + y^2} - D$$

The obtained interferogram on the film, which is located at a distance D from the point that we are trying to record a hologram of, is shown in figure 8.17.

8.3.2.2 Reading of the hologram

Let us now remove the point from the scene, and illuminate the recorded hologram with the same reference beam as the one we used for the recording step. The intensity distribution is actually imprinted on the film that is now used as a diffraction mask. As we saw in section 6.1, in the frame of Fraunhofer diffraction, the distribution of the electric field in a plane after the diffraction mask is the Fourier transform of the mask aperture function. The conditions for being in the Fraunhofer configuration are detailed in chapter 2. Here the mask aperture function is proportional to $I_{\text{point}+\text{ref}}$, given by expression (8.18).

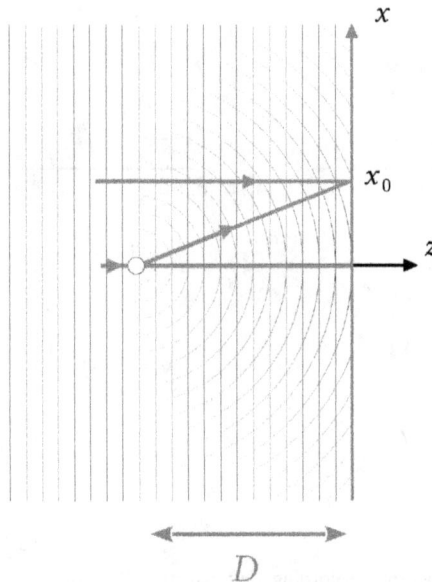

Figure 8.16. Drawing of the principle of holography for a single point in space, represented in 2D. The situation in 3D can be imagined by rotating the drawing around the z-axis.

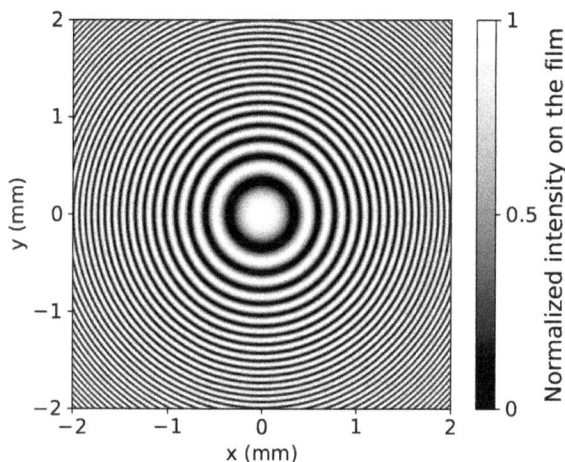

Figure 8.17. Intensity distribution on the holographic film during the recording of the hologram of a single point in space, located at a distance $D = 0.2$ m from the film, for a reference beam wavelength of 600 nm.

$$\mathcal{F}\{T(x, y)\} = \mathcal{F}\left(\alpha + \beta \cos\left(\frac{2\pi}{\lambda_0}(\sqrt{D^2 + x^2 + y^2} - D)\right)\right)$$

To simplify even further the situation, we can consider a plane slice of the situation: indeed, we can see in figure 8.16 that the problem displays an invariance by rotation around the (Oz) axis. Thus we can write that the path difference is:

$$\Delta\delta = \sqrt{D^2 + x^2} - D$$

$$\mathcal{F}\{T(x)\} = \mathcal{F}\left(\alpha + \beta \cos\left(\frac{2\pi}{\lambda_0}(\sqrt{D^2 + x^2} - D)\right)\right)$$

The transmission function evolves with x in such a way that we can approximate it, at each point on the film plane, by a sinusoidal function whose spatial frequency depends on the position of that point. Intuitively, each portion of the film then behaves like a sinusoidal diffraction grating, producing three emerging beams, with angles determined by the local spatial frequency of the transmission function around the point of incidence. To make this intuition precise and mathematically rigorous, we now perform a Taylor expansion of the cosine argument around any value x_0 of the mask. The Taylor expansion formula is given in paragraph (1.3). Applied to $\Delta\delta$ around $x = x_0$, we get:

$$\Delta\delta(x \simeq x_0) \simeq \Delta\delta(x_0) + \left.\frac{\partial\Delta\delta}{\partial x}\right|_{x=x_0} (x - x_0) \text{ with } \frac{\partial\Delta\delta}{\partial x} = \frac{x}{\sqrt{D^2 + x^2}}$$

This new expression for $\Delta\delta$ depends linearly on x:

$$\Delta\delta(x \simeq x_0) \simeq \left(\sqrt{D^2 + x_0^2} - D - \frac{x_0^2}{\sqrt{D^2 + x_0^2}}\right) + x\frac{x_0}{\sqrt{D^2 + x_0^2}}$$

In order to avoid carrying on huge terms, let us write the argument in the cosine as $\Delta\delta(x \simeq x_0) \simeq \lambda_0\,(\gamma + \kappa x)$. The transmission function of the holographic film is thus equal to:

$$T(x \simeq x_0) = \alpha + \beta(\cos(2\pi\gamma)\cos(2\pi\kappa x) - \sin(2\pi\gamma)\sin(2\pi\kappa x))$$

where:

$$\gamma = \frac{1}{\lambda_0}\left(\sqrt{D^2 + x_0^2} - D - \frac{x_0^2}{\sqrt{D^2 + x_0^2}}\right)$$

and:

$$\kappa = \frac{1}{\lambda_0}\frac{x_0}{\sqrt{D^2 + x_0^2}} \tag{8.19}$$

and where we used that $\cos(a + b) = \cos a \cos b - \sin a \sin b$ (see equation (1.1)). These last operations make it easier to take the Fourier transform of the aperture function. Using the linearity theorem, we get that:

$$\mathcal{F}\{T(x)\} = \mathcal{F}\{\alpha\} + \beta\cos(2\pi\gamma)\mathcal{F}\{\cos(2\pi\kappa x)\} - \beta\sin(2\pi\gamma)\mathcal{F}\{\sin(2\pi\kappa x)\}$$

We know the Fourier transform of the cosine and sine functions from section 3.5.1. We get:

$$\mathcal{F}\{T(x)\} = \alpha\delta(\xi) + \frac{\beta}{2}\cos(2\pi\gamma)(\delta(\xi - \kappa) + \delta(\xi + \kappa))$$
$$+ i\frac{\beta}{2}\sin(2\pi\gamma)(\delta(\xi - \kappa) - \delta(\xi + \kappa))$$

Now let us determine what the light intensity distribution looks like from that expression (that is equal to the electric field amplitude distribution). For that purpose, we use equation (2.8) as usual:

$$I \propto \underline{\mathcal{E}} \cdot \underline{\mathcal{E}}^* = (\mathcal{F}\{T(x)\})(\mathcal{F}\{T(x)\})^*$$
$$I \propto \left(\left(\alpha\delta(\xi) + \frac{\beta}{2}\cos(2\pi\gamma)(\delta(\xi - \kappa) + \delta(\xi + \kappa))\right)^2\right.$$
$$\left. + \left(\frac{\beta}{2}\sin(2\pi\gamma)(\delta(\xi - \kappa) - \delta(\xi + \kappa))\right)^2\right)$$

Using that $\delta^2(x) = \delta(x)$ and that:

$$\delta(a)\delta(b) = \begin{cases} \delta(a) \text{ if } a = b \\ 0 \text{ else} \end{cases}$$

one gets:

$$I \propto \alpha^2\delta(\xi) + \frac{\beta^2}{4}\cos^2(2\pi\gamma)(\delta(\xi - \kappa) + \delta(\xi + \kappa))$$
$$+ \frac{\beta^2}{4}\sin^2(2\pi\gamma)(\delta(\xi - \kappa) + \delta(\xi + \kappa))$$
$$I \propto \alpha^2\delta(\xi) + \frac{\beta^2}{4}(\delta(\xi - \kappa) + \delta(\xi + \kappa))$$

where we used that $\cos^2 x + \sin^2 x = 1$ and ignored the special case $\kappa = 0$ where the viewer looks along the optical axis.

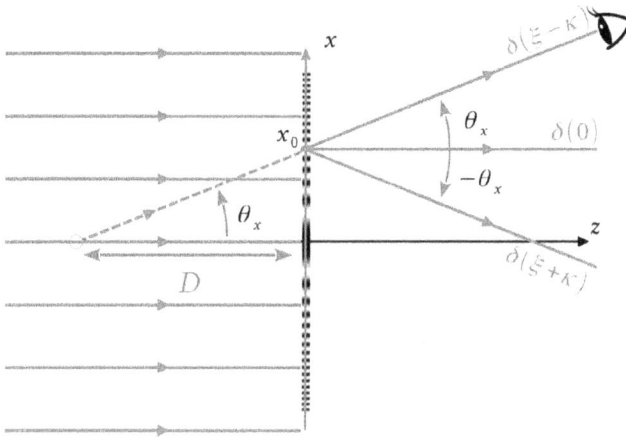

Figure 8.18. Holographic film reading for a single point, seen from the side.

From this expression of the intensity, we see that three beams emerge from the holographic film. One for $\xi = 0$ where we recall that $2\pi\xi = k_x = k \sin \theta_x$. That one peak thus corresponds to an emerging angle of zero. It means that this is the reference beam crossing the mask without being deviated. To eliminate that beam that would be annoying for an observer to see the hologram, that reference beam is usually tilted with respect to the optical axis (Oz). The two other beams emerge with the following values for ξ: $\xi = \pm\kappa$, that correspond to $\sin \theta_x = \pm\lambda_0\kappa$. Replacing κ with its expression, see equation (8.19), we have that:

$$\sin \theta_x = \pm\frac{x_0}{\sqrt{D^2 + x_0^2}}$$

which corresponds to the angle between the optical axis and a virtual light ray that would come from the position where the recorded single point was during the exposition of the holographic film! The situation is displayed in figure 8.18. Now from the information that our brain gets from both our eyes, it locates the point at two different positions on the holographic film (one for each eye). The image gets reconstructed in our brain as a point floating in space at the position where the actual point used to be during the recording.

8.4 Problems

Problem 8.1 Due to the finite extent of the movable mirror path, the Fourier Integral is not evaluated between $-\infty$ and ∞ but rather between $-L$ and L (where $2L$ is the total mirror path). This limitation results in a finite spectral resolution $\Delta\xi$ of the spectrometer, given by $\Delta\xi = \frac{1}{2L}$. Demonstrate this relation. A typical mirror path is 5 cm, what would be the spectral resolution?

Problem 8.2 An optical intensity distribution is given by $f(x) = \cos(6\pi x)$ sampled with a frequency of $\xi_0 = 5$.

(a) Will aliasing occur? Explain why or why not.

(b) Sketch the Fourier transform $F(\xi)$ of the original signal $f(x)$.

(c) Sketch the Fourier transform $F(\xi)$ of the sampled signal $f(nL)$. Make a graphic of the Fourier transform of the sampled function.

(d) Suppose a low-pass filter (rectangular function) with a width of $\frac{\xi_0}{2} = 2.5$ is applied to the Fourier transform of the sampled function. Represent the result.

(e) What is the obtained function if the inverse Fourier transform is applied to the resulting function in (d).

Further Reading

[1] Lathi B P and Green R 2017 *Linear Systems and Signals* 3rd edn (Oxford: Oxford University Press)

[2] Proakis J G and Manolakis D G 2006 *Digital Signal Processing: Principles, Algorithms, and Applications* 4th edn (Upper Saddle River, NJ: Pearson)

[3] Oppenheim A V and Willsky A S 1996 *Signals and Systems* 2nd edn (Upper Saddle River, NJ: Prentice-Hall)

References

[1] Françon M 1987 *Holographie* 2nd edn (Paris: Masson)

[2] Goodman J W 2017 *Introduction to Fourier Optics* 4th edn (New York: W.H. Freeman)

[3] Sanderson G 2024 How are holograms possible? *YouTube video* https://www.youtube.com/watch?v=EmKQsSDlaa4

www.ingramcontent.com/pod-product-compliance
Lightning Source LLC
Chambersburg PA
CBHW080548220326
41599CB00032B/6409